Naha Giant Tug of War
Naha City, Okinawa Prefecture
200m

Echigosekikawa Great Serpent Festival
Sekikawa Village, Iwafune-gun,
Niigata Prefecture
83m

Comparing Lengths

Seto Bridge
Kurashiki City, Okayama Prefecture
9368m

Irabu Bridge
Miyakojima City, Okinawa Prefecture
3540m

Horai Bridge
Shimada City, Shizuoka Prefecture
897m

蓬莱橋

Choshi Bridge
Choshi City, Chiba Prefecture
1209m

230m

290m

Kohun is a
mounded tomb of
ancient Japan.

Nakatsuhime-no-mikoto-ryo Kofun (left)
Ingyo-tenno-ryo Kofun (right)
Fujiidera City, Osaka Prefecture

JN119571

Table of Contents

Nanami

Hiroto

Reflect · Connect ···································· 82

Daiki

Let's learn mathematics together.

Grade 3 Vol.2

Yui

3

① Thinking Competency

Competency to think the same or similar way
Competency to find what you have learned before and think in the same or similar way

Competency to find rules
Competency to analyze various numbers and expressions and investigate any rules

Competency to explain the reason
Competency to explain the reason why, based on learned rules and important ideas

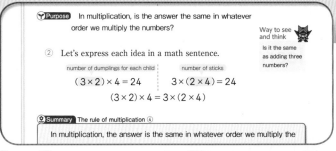

Purpose In multiplication, is the answer the same in whatever order we multiply the numbers?

Way to see and think
Is it the same as adding three numbers?

② Let's express each idea in a math sentence.

number of dumplings for each child
$(3 \times 2) \times 4 = 24$

number of sticks
$3 \times (2 \times 4) = 24$

$(3 \times 2) \times 4 = 3 \times (2 \times 4)$

Summary The rule of multiplication ④

In multiplication, the answer is the same in whatever order we multiply the

② Judgement Competency

Competency to find mistakes
Competency to find the over generalization of conventions and rules

Competency to find the correct
Competency to find the correct way by comparing the numbers or the quantities

Competency to compare ideas and way of thinking
Competency to find the same or different ways of thinking by comparing friends' ideas and your own ideas

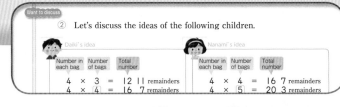

Want to discuss
② Let's discuss the ideas of the following children.

Daiki's idea

Number in each bag		Number of bags		Total number	
4	×	3	=	12	11 remainders
4	×	4	=	16	7 remainders

Nanami's idea

Number in each bag		Number of bags		Total number	
4	×	4	=	16	7 remainders
4	×	5	=	20	3 remainders

③ Representation Competency

Competency to represent sentences with pictures or expressions
Competency to read problem sentences, draw diagrams, and represent them using expressions

Competency to represent data in graphs or tables
Competency to express the explored data in tables or graphs in simpler and easier ways

Competency to communicate with your friends
Competency to communicate your ideas to friends in simpler and easier manners and to write notes

How much was the total cost ?

cracker 215 yen
chocolate 143 yen

Way to see and think

Remember diagram fo addition yo have learne the 2nd gra

① Let's draw a diagram.

Let's find monsters.

Monsters
which represent ways of thinking in mathematics

Setting the unit.
Once you have decided one unit, you can represent how many.

Unit

If you try to separate...
Decomposing numbers by place value and dividing figures make it easier to think about problems.

Separate

If you represent in other way...
If you represent in other expression, diagram, table, etc., it is easier to understand.

Other way

Can you do the same or similar way?
If you find something the same or similar, then you can understand.

Looks same

You wonder why?
Why this happens? If you communicate the reasons in order, it will be easier to understand for others.

Why

If you try to arrange...
You can compare if you align the number place and align the unit.

Align

If you try to summarize...
It is easier to understand if you put the numbers in groups of 10 or summarize in a table or graph.

Summarize

If you try to change the number or figure...
If you try to change the problem a little, you can better understand the problem or find a new problem.

Change

Is there a rule?
If you examine a few examples, then you can find out whether there is a rule.

Rule

Ways to think learned in the 2nd grade

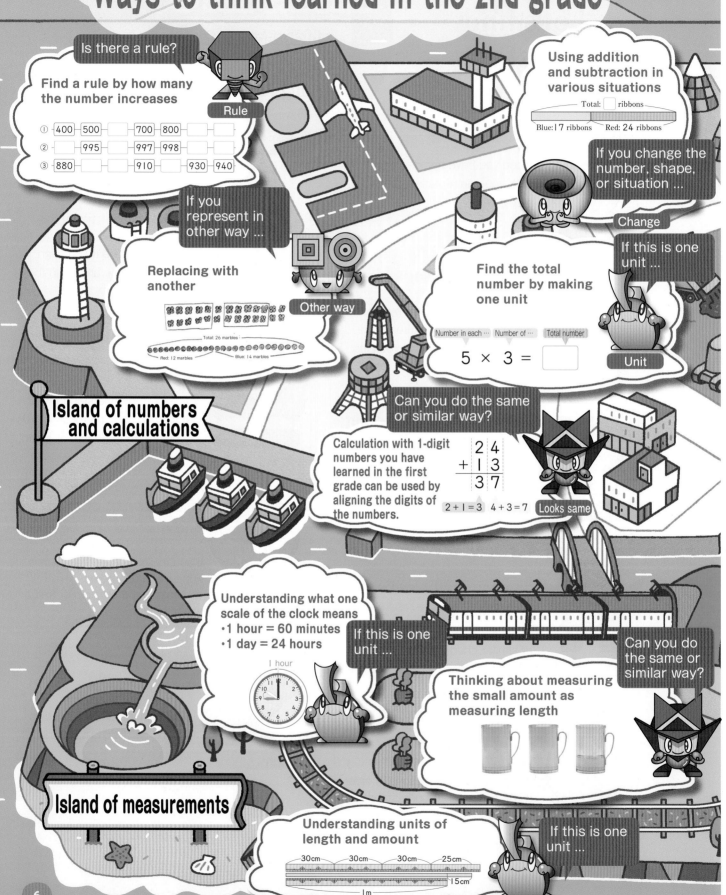

Is there a rule?

Rule

Find a rule by how many the number increases

① 400 — 500 — ☐ — 700 — 800 — ☐ — ☐
② ☐ — 995 — ☐ — 997 — 998 — ☐ — ☐
③ 880 — ☐ — ☐ — 910 — ☐ — 930 — 940

If you represent in other way ...

Other way

Replacing with another

Total: 26 marbles
Red: 12 marbles   Blue: 14 marbles

Island of numbers and calculations

Using addition and subtraction in various situations

Total: ☐ ribbons
Blue: 17 ribbons   Red: 24 ribbons

If you change the number, shape, or situation ...

Change

If this is one unit ...

Find the total number by making one unit

Number in each ··· Number of ··· Total number

$5 \times 3 = $ ☐

Unit

Can you do the same or similar way?

Calculation with 1-digit numbers you have learned in the first grade can be used by aligning the digits of the numbers.

$$\begin{array}{r} 2\ 4 \\ +\ 1\ 3 \\ \hline 3\ 7 \end{array}$$

$2 + 1 = 3$ $4 + 3 = 7$

Looks same

Understanding what one scale of the clock means
• 1 hour = 60 minutes
• 1 day = 24 hours

1 hour

If this is one unit ...

Thinking about measuring the small amount as measuring length

Can you do the same or similar way?

Island of measurements

Understanding units of length and amount

30cm   30cm   30cm   25cm
15cm
1m

If this is one unit ...

Let's progress through 3 ways of learning.

⭐1 Self directed learning: Learning on your own initiative.

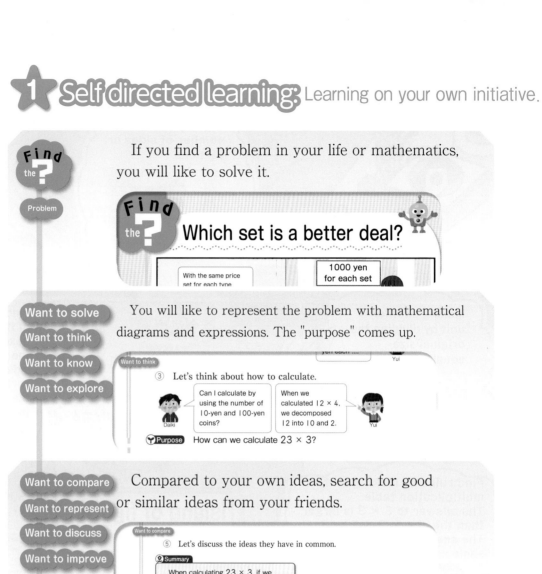

Find the ?

Problem

If you find a problem in your life or mathematics, you will like to solve it.

Find the ? Which set is a better deal?

With the same price set for each type

1000 yen for each set

Want to solve
Want to think
Want to know
Want to explore

You will like to represent the problem with mathematical diagrams and expressions. The "purpose" comes up.

Want to think

③ Let's think about how to calculate.

Daiki: Can I calculate by using the number of 10-yen and 100-yen coins?

Yui: When we calculated 12×4, we decomposed 12 into 10 and 2.

🎯 Purpose: How can we calculate 23×3?

Want to compare
Want to represent
Want to discuss
Want to improve
Want to communicate

Compared to your own ideas, search for good or similar ideas from your friends.

Want to compare

⑤ Let's discuss the ideas they have in common.

🔍 Summary

When calculating 23×3, if we decompose 23 into 20 and 3, we can calculate by using the multiplication table.

$23 \times 3 \begin{cases} 3 \times 3 = 9 \end{cases}$

Want to connect

Want to confirm

Make sure to confirm that the "summary" can be used for other problems.

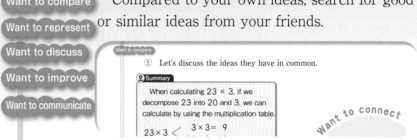

Want to confirm

2 Divide 10 dL of juice equally among 5 children.
How many dL of juice does each one get?

Way to see and think
Which row in the multiplication

Want to try

When you can do it, you will want to have more problems.

Want to try

3 Which row in the multiplication table should you use to do the following divisions? Let's find the answers.

① $8 \div 2$ ② $21 \div 7$ ③ $72 \div 9$ ④ $28 \div 4$
⑤ $20 \div 5$ ⑥ $56 \div 8$ ⑦ $21 \div 3$ ⑧ $54 \div 6$

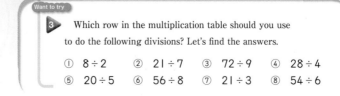

⭐2 Dialogue learning: Learning together with friends.

As learning progresses, you will want to know what your friends are thinking. Also, you will like to share your own ideas with your friends. Let's discuss with the person next to you, in groups, or with the whole class.

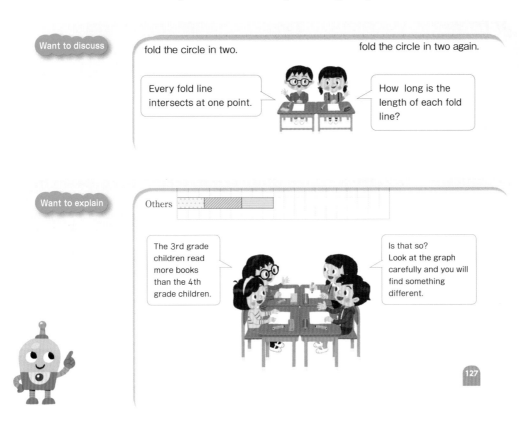

Want to discuss

fold the circle in two.

fold the circle in two again.

Every fold line intersects at one point.

How long is the length of each fold line?

Want to explain

Others

The 3rd grade children read more books than the 4th grade children.

Is that so? Look at the graph carefully and you will find something different.

127

⭐3 Deep learning: Usefulness and efficiency of what you learned.

Let's cherish the feeling "I want to know more." and "Can I do this in another case?" Let's deepen learning in life and mathematics.

351 + 97 + 49

Can make a math expression and find the answer.

Let's deepen.

Can we utilize calculations of larger numbers in life?

Daiki

79

Utilize in life.

Deepen.

Let's go shoppi

Riku went to some stores to sh

He brought money and a shoppi

Want to connect

Can I calculate 213 × 3 in vertical form ?

Daiki

Can you find the answer by using the multiplication table?

There are many chocolates.

1

4 × 6 in this box ....

2

3 × 6 in this box ....

So, how many are there altogether ....

3

We can get the total number by calculating 7 × 6.

....

4

Problem　Let's think about why we can get the answer by calculating 7 × 6.

1 Multiplication
Let's find the rules of multiplication and apply them beyond 9 × 9.

Want to find

Activity

1 Let's write what you have found from the multiplication table in diagrams or math expressions.

Multiplier

	1	2	3	4	5	6	7	8	9
1									
2									
3									
4									
5									
6						42			
7					42				
8									
9									

Multiplicand

Way to see and think

In multiplication, if the multiplier increases by 1, the answer increases by the multiplicand.

① Write all the answers in the table above.

Hiroto

If the multiplier increases by 1, the answer ....

There are many answers that are the same.

Nanami

Way to see and think

What rules can you find in the multiplication table?

Purpose What rules are there in multiplication?

② Hiroto and Yui found the following rules in the multiplication table. Let's explain what rules they found.

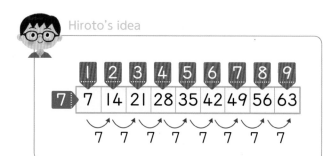

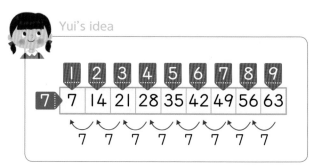

③ When 7×6 is expressed in other math expressions by using the ideas of Hiroto and Yui, each of them can be written as the following. Let's fill in the ☐ with numbers.

Way to see and think

Replace it by a diagram using ●.

$$7 \times 6$$
$$7 \times 5$$

Hiroto $7 \times 6 = 7 \times 5 + \boxed{}$

Yui $7 \times 6 = 7 \times 7 - \boxed{}$

The symbol "=" is called the **equality sign**. It is not only used for writing the answer to any operation, but also for showing that the expressions or the size of numbers on the left side and the right side are equal.

⚙ **Summary** The rule of multiplication ①

In multiplication, if the multiplier increases by 1, the answer increases by the multiplicand. Also, if the multiplier decreases by 1, the answer decreases by the multiplicand.

1 Let's fill in the ☐ with numbers.

① 4×6 is larger than 4×5 by ☐.

② 5×8 is smaller than 5×9 by ☐.

③ $7 \times 7 = 7 \times$ ☐ $+ 7$ ④ $3 \times$ ☐ $= 3 \times 7 - 3$

 Nanami found the following rule in the multiplication table.

Let's explain what rule she found.

Nanami's idea

 The answer to 7×6 is the same as the one to ☐ × ☐.

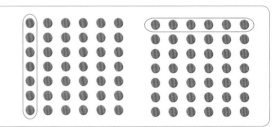

① When Nanami's idea is expressed in a math sentence, it can be written as the following. Let's fill in the ☐ with a number.

$7 \times 6 = 6 \times$ ☐

Summary The rule of multiplication ②

In multiplication, the answer is the same even if the order of the multiplicand and the multiplier is switched.

 2 Let's fill in the ☐ with numbers.

① $8 \times 7 =$ ☐ $\times 8$ ② $9 \times$ ☐ $= 3 \times 9$

Activity

3 In 7 × 6, let's think about what will happen to the answer if you decompose the multiplicand or the multiplier.

 Daiki: Can I think about it by decomposing 7 × 6 into 4 × 6 and 3 × 6?

 Nanami: Can I calculate by decomposing the multiplier?

Purpose Will we get the same answer if we decompose the multiplicand or the multiplier?

① Let's explain the ideas of Daiki and Nanami.

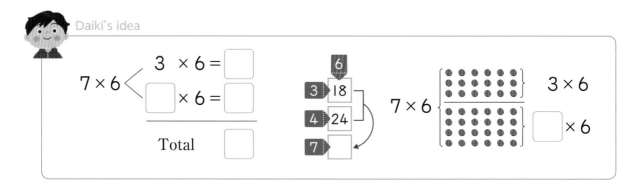

Daiki's idea

7×6 ⟨ $3 \times 6 = \square$ $\square \times 6 = \square$

Total $\square$

7×6 : 3 ▸ 18, 4 ▸ 24, 7 ▸ $\square$ 6

7×6 ⟨ 3×6 $\square \times 6$

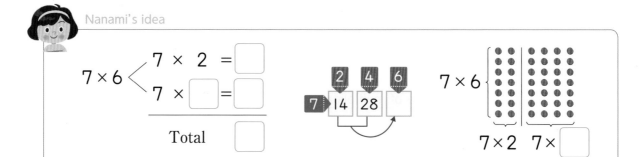

Nanami's idea

7×6 ⟨ $7 \times 2 = \square$ $7 \times \square = \square$

Total $\square$

7 ▸ 14, 28, $\square$ 2 4 6

7×6 ⟨ 7×2 $7 \times \square$

Summary The rule of multiplication ③

In multiplication, we get the same answer even if we decompose the multiplicand or the multiplier.

② Hiroto thought about the calculation of 7×6 as the following. Let's fill in the ☐ with numbers and talk about Hiroto's idea.

Hiroto's idea

$$7 \times 6 \begin{cases} 7 \times 5 \quad \boxed{} \\ 7 \times \boxed{} = \boxed{} \end{cases}$$

Total ☐

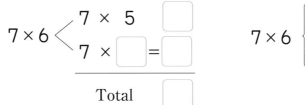

$$7 \times 6 \begin{cases} \end{cases}$$

7×5 $\quad 7 \times \boxed{}$

If I express this in a math sentence,

$$7 \times 6 = 7 \times 5 + \boxed{}$$

So, it is the same as the rule of multiplication ① to calculate by decomposing the multiplier into 1 and some in the rule of multiplication ③.

③ Let's think about whether other calculations can be expressed in a math sentence by decomposing the multiplier into 1 and some.

3 Let's fill in the ☐ with numbers.

① $7 \times 9 \begin{cases} 5 \times 9 = \boxed{} \\ \boxed{} \times 9 = \boxed{} \end{cases}$

Total ☐

② $4 \times 8 \begin{cases} 4 \times 2 = \boxed{} \\ 4 \times \boxed{} = \boxed{} \end{cases}$

Total ☐

4 Each child gets 2 sticks of 3 dumplings. How many dumplings are needed for 4 children?

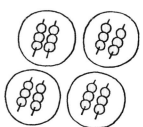

① Yui and Hiroto found the answer as shown below.

Let's explain how Yui and Hiroto thought.

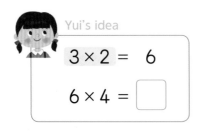

Yui's idea

$$3 \times 2 = 6$$

$$6 \times 4 = \boxed{}$$

Hiroto's idea

$$2 \times 4 = 8$$

$$3 \times 8 = \boxed{}$$

Purpose In multiplication, is the answer the same in whatever order we multiply the numbers?

② Let's express each idea in a math sentence.

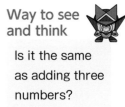

Way to see and think

Is it the same as adding three numbers?

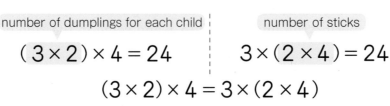

number of dumplings for each child

$$(3 \times 2) \times 4 = 24$$

number of sticks

$$3 \times (2 \times 4) = 24$$

$$(3 \times 2) \times 4 = 3 \times (2 \times 4)$$

Summary The rule of multiplication ④

In multiplication, the answer is the same in whatever order we multiply the numbers.

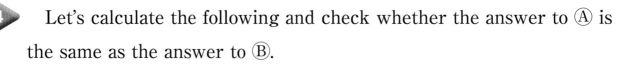

The math expression inside the ( ) should be done first.

$$(3 \times 2) \times 4 = 3 \times (2 \times 4)$$

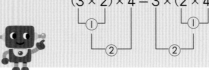

4 ▶ Let's calculate the following and check whether the answer to Ⓐ is the same as the answer to Ⓑ.

① Ⓐ $(2 \times 3) \times 3$ Ⓑ $2 \times (3 \times 3)$

② Ⓐ $(2 \times 4) \times 2$ Ⓑ $2 \times (4 \times 2)$

Notebook for thinking

Let's write the ideas and doubts that you had.

April 18

> Each child gets 2 sticks of 3 dumplings. How many dumplings are needed for 4 children?

Purpose : Let's think about how to get the total number.

〈My idea〉
Each child gets 2 sticks of 3 dumplings. So the math expression becomes 3×2.
There are 4 children and each child gets ○ dumplings.
So, ○ $\times$ 4 is the total number.
How can I write this in easier ways to understand?

〈Reflect〉
· Only based on the math sentence of
  $(3 \times 2) \times 4 = 3 \times (2 \times 4)$, is it true that the answer is the same in whatever order we multiply the numbers?
· Is it expressed as $3 \times 4 \times 2$?

〈Haruto's idea〉
Find the number of sticks first.
Then, there are 8 sticks of 3 dumplings,
Math Sentence : $2 \times 4 = 8$ $3 \times 8 = 24$
            Answer : 24 dumplings
Try to express in one math sentence.
$3 \times (2 \times 4) = 24$

Write today's date.

Write the problem of the day that you must know.

Let's learn with the purpose.

Write your ideas frankly.

As for reflection, the following must be written:
· what you understood,
· interesting thoughts
· doubts you had
· how you felt about classmates' ideas
· what you want to do more

Want to try

Point Scoring Game

You are given 10 marbles to shoot at the target. The points that you can get depends on where the marbles land. The one who gets the highest score wins.

Want to explore

1 Sota's result of the game is shown on the right. Let's see his score.

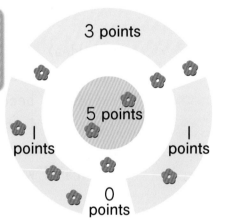

Points	5	3	1	0	Total
Number of marbles					
Score					

① Let's write the math expressions to find the total points.

5-point : 2 marbles ·········· 5 × ☐

3-point : 0 marbles ·········· 3 × ☐

1-point : 4 marbles ·········· 1 × ☐

0-point : 4 marbles ·········· 0 × ☐

points   number of marbles

Nanami

0 × 4 is
0 + 0 + 0 + 0,
so ....

Can I use any rules of multiplication?

Daiki

⊻ Purpose When the multiplier or multiplicand is 0 in the multiplication, what is the answer to it?

② Let's find the score for 3-point by using the row of 3 in the multiplication table.

$3 \times 4 = 12$
$3 \times 3 = 9$
$3 \times 2 = 6$
$3 \times 1 = 3$
$3 \times 0 = \boxed{}$

decreased by $\boxed{}$

Way to see and think

Thinking by using the rule of multiplication ①.

③ Let's find the score for 0-point in easier ways.

$0 \times 4 = \boxed{}$

④ Let's complete the table on the left page and find the total score of Sota.

 In the point scoring game, what does the math expression 0×0 mean?

⚘ Summary

Any number multiplied by 0 equals 0. $3 \times 0 = 0$
Also, 0 multiplied by any number equals 0. $0 \times 4 = 0$
        $0 \times 0 = 0$

 Let's calculate the following.

① 6×0 ② 4×0 ③ 0×7 ④ 0×1

 In the multiplication table on page 141, let's write the answers in the multiplicands of 0 and in the multipliers of 0.

❸ Multiplication with 10

Activity

Want to represent

1 How many stickers are there in total?

① Let's write **2** math expressions to find the number of stickers.

▢ × ▢   ▢ × ▢

Want to think

② Let's think about how to find the answer to $5 × 10$.

Way to see and think

Thinking by using the rule of multiplication ① and ③.

Hiroto's idea

$5 × 10 = 5 × 9 + ▢$

Nanami's idea

$5 × 10 \begin{cases} 5 × 2 = 10 \\ 5 × ▢ = ▢ \end{cases}$

Total ▢

③ Let's find the answer to $10 × 5$.

Way to see and think

Can you use the rule of multiplication ② or ③?

If I decompose 10 into 7 and 3, I will get $7 × 5$ and $3 × 5$ ....

Yui

$10 × 5 = 5 × 10$ So ....

Daiki

Want to discuss

▶**1** Let's discuss how to find the answer to $10 × 10$.

Want to confirm

▶**2** There are **7** children and each one is given **10** stickers. How many stickers are needed altogether?

▶**3** Let's calculate the following.

① $6 × 10$ ② $8 × 10$ ③ $10 × 4$ ④ $10 × 9$

▶**4** In the multiplication table on page **141**, let's write the answers in the multiplicands of **10** and in the multipliers of **10**.

What you can do now

☐ Understanding the rule of multiplication ①.

I Let's fill in the ☐ with numbers or expressions.

① In the row of 5, if the multiplier increases by 1, the answer increases by ☐.

② In the row of 9, if the multiplier decreases by 1, the answer decreases by ☐.

③ 3×9 is larger than 3×8 by ☐.

Show in a math sentence. $3 \times 9 =$ ☐

④ 4×3 is smaller than 4×4 by ☐.

Show in a math sentence. $4 \times 3 =$ ☐

☐ Understanding the rule of multiplication ②.

2 Let's fill in the ☐ with numbers.

① $3 \times 8 = 8 \times$ ☐   ② $4 \times$ ☐ $= 6 \times 4$

☐ Understanding the rule of multiplication ③.

3 Let's fill in the ☐ with numbers.

① 8×7 〈 $8 \times 3 =$ ☐
                $8 \times$ ☐ $=$ ☐

Total ☐

② 9×6 〈 $5 \times 6 =$ ☐
                ☐ $\times 6 =$ ☐

Total ☐

☐ Understanding the rule of multiplication ④.

4 Let's fill in the ☐ with numbers.

① $(3 \times 3) \times 2 = 3 \times ($ ☐ $\times 2)$ ② $(7 \times 2) \times 4 = 7 \times (2 \times$ ☐ $)$

☐ Can multiply with 0 and 10.

5 Let's calculate the following.

① 0×9 ② 8×0 ③ 0×0

④ 2×10 ⑤ 10×6 ⑥ 10×2

Supplementary problems ▶ p.128

Usefulness and efficiency of learning

1 Parts of the multiplication table are shown below.

Let's fill in the numbered spaces with answers.

☐ Understanding the rule of multiplication.

①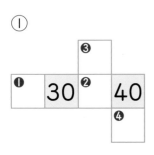

②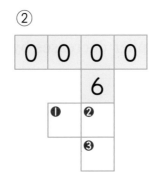

③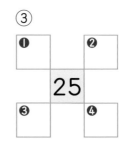

2 There are 3 boxes of chocolates that contain 10 chocolates each and 10 boxes that contain 6 chocolates each. How many chocolates are there altogether?

☐ Can multiply with 10.

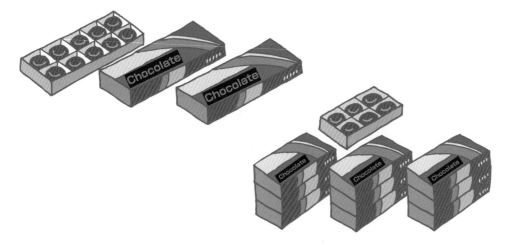

3 Let's think about how to calculate 4 × 12 by using the rules of multiplication.

☐ Understanding the rule of multiplication.

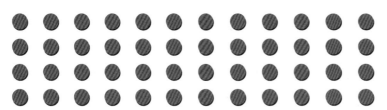

What time did you arrive?

Let's start a field trip.

We arrived at the park.

It took 30 minutes from our school.

 Problem What time did they arrive at the park?

23

Let's find time and duration and utilize in life.

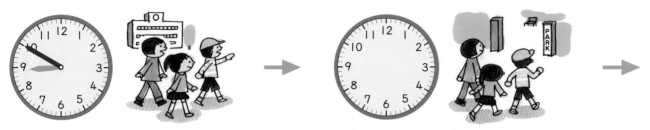

The time they left school   The time they arrived at the park

Want to know What time past

1
Daiki and his friends left school at 8:50 a.m. It took 30 minutes to arrive at the park on foot. What time did they arrive at the park?

Hiroto

Let's think about it by using pictures of a clock or a line of time.

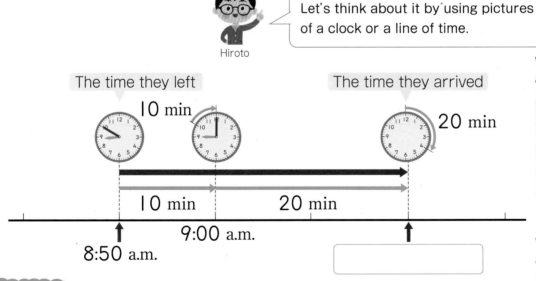

The time they left   The time they arrived

10 min   20 min

10 min   20 min

9:00 a.m.
8:50 a.m.

Way to see and think

One scale represents 10 minutes by replacing a clock by a line.

Way to see and think

It is easy to understand by replacing duration or time by a line of number.

Want to confirm

 1 Let's find the following time.

① The time 50 minutes past 9:30 a.m.

② The time 20 minutes past 1:50 a.m.

③ The time 1 hour 10 minutes past 2:40 p.m.

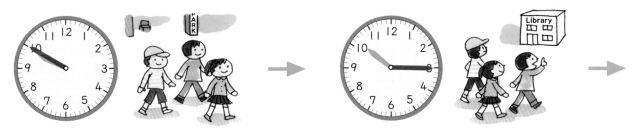

The time they left the park   The time they arrived at the library

Want to know How long it took

2 They left the park at 9:50 a.m. and arrived at the library at 10:15 a.m. How long did it take from the park to the library?

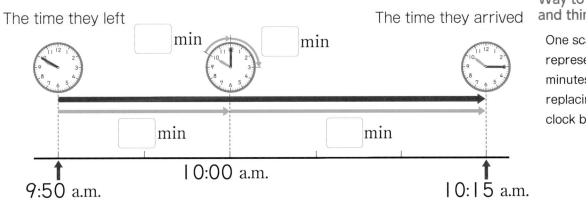

The time they left   The time they arrived

☐ min   ☐ min

☐ min   ☐ min

10:00 a.m.

9:50 a.m.   10:15 a.m.

Way to see and think

One scale represents 5 minutes by replacing a clock by a line.

The question of "how long" is answered by duration or period of time.

Want to confirm

3 Let's find the following duration.

① The duration from 9:30 a.m. to 10:20 a.m.

② The duration from 3:10 p.m. to 4:50 p.m.

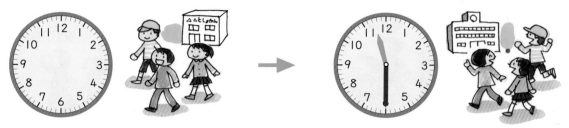

The time they left the library   The time they arrived at school

2   They left the library and walked to school for 40 minutes. They arrived there at 11:30 a.m. What time did they leave the library?

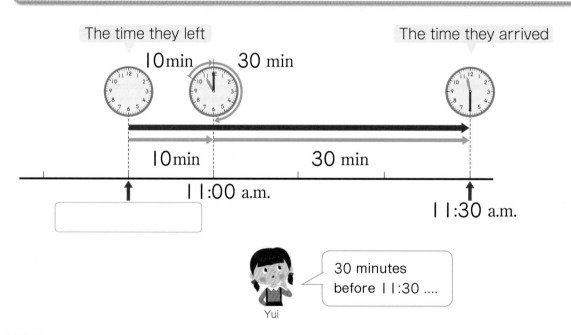

Yui
30 minutes before 11:30 ....

4   Let's find the following time.

①   The time 50 minutes before 11:20 a.m.

②   The time 1 hour 20 minutes before 2:20 p.m.

5 They were in the park for 30 minutes and in the library for 35 minutes. How long was the sum of the duration they were in the park and in the library?

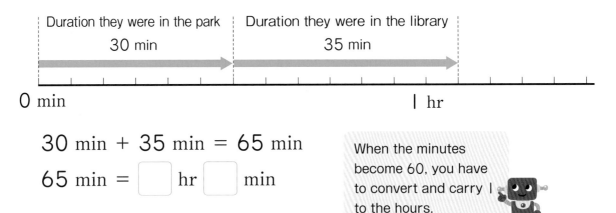

Duration they were in the park | Duration they were in the library
30 min | 35 min

0 min   1 hr

30 min + 35 min = 65 min

65 min = ☐ hr ☐ min

When the minutes become 60, you have to convert and carry 1 to the hours.

 6 The duration they walked on their field trip is shown on the right. How long did they walk altogether?

from the school to the park   30 min
from the park to the library   25 min
from the library to the school 40 min

 7 They left school at 8:50 a.m. and came back to school at 11:30 a.m. Let's find the duration they were out of school by writing an expression.

11:30 ☐ 8:50

Want to connect

Can I calculate time by using vertical form we have learned in the 2nd grade?

Hiroto

3 Let's think about how to calculate the expression
11:30 – 8:50 you found in the previous page by
using the vertical form shown on the right.

```
     Hr   Min
     11   30
  -   8   50
  _____
```

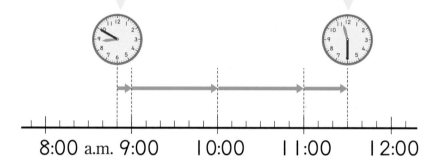

The time they left school   The time they came back to school

8:00 a.m. 9:00   10:00   11:00   12:00

How to calculate time and duration in vertical form

```
   Hr      Min
   11      30
 -  8      50
 _____
```

Align the
"hours" and
"minutes."

→

```
   Hr      Min
   10      60
   11      30
 -  8      50
 _____
           40
```

1 hour is converted to 60 minutes
and borrowed to the minutes.
60 + 30 = 90
90 − 50 = 40

→

```
   Hr      Min
   10      60
   11      30
 -  8      50
 _____
    2      40
```

Because 1 hour is
borrowed,
10 − 8 = 2
So, 2 hours 40 minutes

---

8 What time is 2 hours 50 minutes past 8:30 p.m.?

```
   Hr      Min
    8      30
 +  2      50
 _____
```

When the minutes become 60,
they are converted to 1 hour
and carry to the hours.

What you can do now

☐ Can find the time past.

1 Let's find the following time.

① The time 30 minutes past 4:10 a.m.

② The time 35 minutes past 3:25 p.m.

③ The time 1 hour 20 minutes past 7:36 p.m.

☐ Can find the time before.

2 Let's find the following time.

① The time 30 minutes before 11:45 a.m.

② The time 45 minutes before 6:30 p.m.

③ The time 1 hour 15 minutes before 9:50 a.m.

☐ Can find the total duration.

3 One Sunday, Yuta read a book for 1 hour 10 minutes in the morning and 45 minutes in the afternoon. Altogether, how long did he read the book on that day?

☐ Can find how long it took.

4 Hikari practiced the piano from 9:30 a.m. to 11:10 a.m. How long did she practice?

Supplementary problems ········► p.129

Usefulness and efficiency of learning

1 Let's find the following time.

☐ Can find the time past.

① The time 20 minutes past 8:18 p.m.

② The time 50 minutes past 10:20 a.m.

③ The time 1 hour 35 minutes past 3:55 p.m.

④ The time 48 minutes past 5:42 a.m.

2 Yuna is going to Yota's house through the front of the post office. It takes 10 minutes to walk from her house to the post office and 15 minutes to walk from the post office to his house.

☐ Can find the time before.

　　In order to arrive at his house at 2:00 p.m., what time does she have to leave home?

☐ Can find the total duration.

3 It takes 2 hours 45 minutes in the morning and 3 hours 30 minutes in the afternoon to complete Kento's school trip. Altogether, how long does it take for the school trip?

☐ Can find how long it took.

4 Sayaka practiced soccer from 2:35 p.m. to 4:20 p.m. How long did she practice?

Let's deepen.

We can make various plans when we know the time and duration for them.

Nanami

Deepen.

Utilize in life.

Let's play at the amusement park!

You don't have to think about the duration it takes to wait or move between attractions.

Want to know

Taiga and his friends are going to the amusement park. They checked the duration it takes for one round of a ride or an event. Look at the table below and answer the following.

Duration it takes for one round of a ride or an event

Ride · Event	Ferris wheel	Go-cart	Haunted House	Roller Coaster	Parade
Duration	10 min	4 min	12 min	3 min	15 min

① It takes **25** minutes from Taiga's house to the amusement park. What time will they arrive at the amusement park if they leave home at **9:40** a.m.?

② Soon after they arrive at the amusement park, they want to first ride in the Ferris wheel once and then play with the Go-cart twice. What time will they finish with the rides?

③ Taiga wants to go to the Haunted House before they have lunch at **11:00** a.m. By what time does he have to go there?

Want to deepen

They want to see the Parade at **1:00** p.m. before leaving the amusement park at **1:30** p.m. Which of the following combinations of rides or events can they do before going home?

ⓐ **1** round each of Ferris wheel, Go-cart, and Roller Coaster

ⓑ **1** round each of Haunted House and Roller Coaster

ⓒ **2** rounds each of Go-cart and Roller Coaster

ⓓ **1** round each of Haunted House and Go-cart

31

Find the ?

Can you share cookies equally?

Problem
· If you divide 12 cookies equally among 4 children, how many cookies does each child get?

· If you divide 12 cookies so that each child gets 4 cookies, how many children share the cookies?

3 Division
Let's think about how to divide things equally.

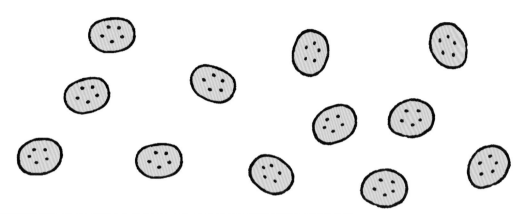

1 Calculation to find the number for each child

Want to solve

1 If you divide 12 cookies equally among 4 children, how many cookies does each child get?

① Let's examine how many cookies each child gets by using blocks.

First, distribute one cookie on each dish.

Hiroto

Purpose If we divide cookies equally, how many cookies does each child get? Can we express the number in a math sentence?

② Let's explain how to divide the cookies as shown below.

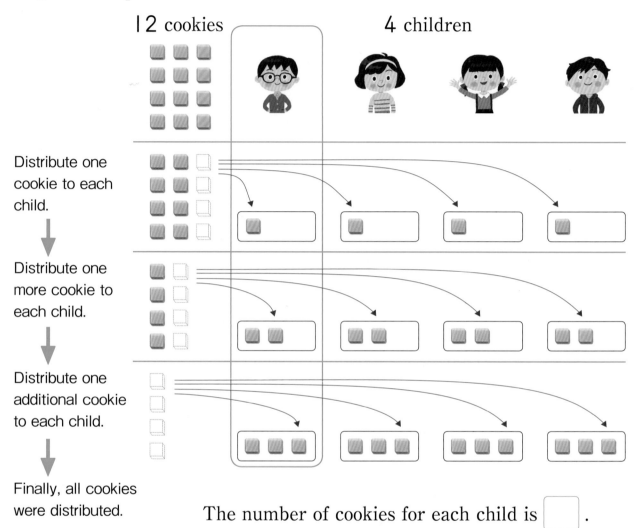

12 cookies   4 children

Distribute one cookie to each child.

Distribute one more cookie to each child.

Distribute one additional cookie to each child.

Finally, all cookies were distributed.

The number of cookies for each child is ⬚.

If you divide the 12 cookies equally among 4 children, each child gets 3. In a math sentence, it can be written as 12 ÷ 4 = 3, and is read as "12 divided by 4 equals 3."

12 ÷ 4 = 3   Answer : 3 cookies

Total number   Number of units   Number for each unit

This kind of operation is called **division.**

In this math sentence, the number of units is the number of children and the number for each unit is the number of cookies for each child.

🌸 **Summary**

When the total number of cookies is equally divided by the number of children, the number of cookies for each child is represented by a division sentence.

Want to confirm

▶1   Let's write a math sentence for each situation below and find the number of blocks given to each child.

① Divide **6** blocks equally among **3** children.

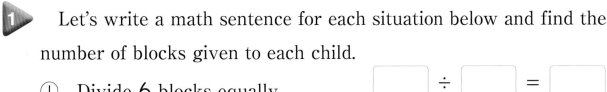

② Divide **15** blocks equally among **5** children.

③ Let's try changing the number of blocks and children.

Want to solve   How to find the answer to a division

2   Divide **15** candies equally among **3** children. How many candies does each child get?

① Let's write a math expression.

② Let's think about how to calculate.

🎯**Purpose**   Let's think about how to find the answer of a division to find the number of candies for each child.

Remember
"Number of candies for each child × Number of children = Total number."

Yui

35

The number of candies for each child is 3.		$3 \times 3 = 9$
The number of candies for each child is 4.		$4 \times 3 = 12$
The number of candies for each child is 5.		$5 \times 3 = 15$

Number for each child   Number of children   Total number

The answer to $15 \div 3$ is the number to be placed in the ☐ for ☐ $\times 3 = 15$.

Math Sentence : $15 \div 3 = 5$ Answer : 5 candies

Way to see and think

Various numbers are placed as the number for each child.

Summary

The answer to $15 \div 3$ can be found by using the row of 3 in the multiplication table.

☐ $\times 3 = 3 \times$ ☐
so,
$3 \times ③ = 9$
$3 \times ④ = 12$
$3 \times ⑤ = ⑮$

Want to confirm

2   Divide $10\,dL$ of juice equally among 5 children.

How many dL of juice does each one get?

Way to see and think

Which row in the multiplication table should you use?

Want to try

3   Which row in the multiplication table should you use to do the following divisions? Let's find the answers.

① $8 \div 2$ ② $21 \div 7$ ③ $72 \div 9$ ④ $28 \div 4$

⑤ $20 \div 5$ ⑥ $56 \div 8$ ⑦ $21 \div 3$ ⑧ $54 \div 6$

3 Look at the picture and make a problem that can be solved by division.

①

8 chocolates

⬜ chocolates are to be divided equally among ⬜ children. How many pieces does each child get?

②

③ Let's make a math problem for the math expression 6 ÷ 2.

There are 6 apples. If 2 children share the apples equally ....

There are 6 sweets and 2 plates ....

4 Let's calculate the following.

① 14 ÷ 2 ② 27 ÷ 9 ③ 40 ÷ 5 ④ 32 ÷ 8

⑤ 18 ÷ 3 ⑥ 42 ÷ 7 ⑦ 48 ÷ 8 ⑧ 12 ÷ 6

Want to solve

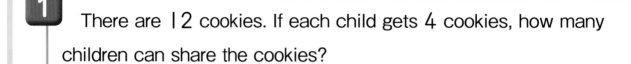

1 There are 12 cookies. If each child gets 4 cookies, how many children can share the cookies?

① Let's examine how many children can share the cookies by using blocks.

Purpose When the cookies are distributed so that each child gets an equal number, how many children can share them?
Can we express the number in a math sentence?

Want to explain

② Let's explain how to divide the cookies as shown below.

 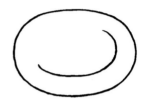

Distribute 4 cookies to one child.

Distribute 4 cookies to one more child.

Distribute 4 cookies to one additional child.

Finally, all cookies were distributed.

☐ children can share the cookies.

If you divide **12** cookies so that each child gets **4** cookies, then **3** children can equally share the cookies. By using a division sentence, it can be written as **12 ÷ 4 = 3**.

$$12 \div 4 = 3$$

12	÷	4	=	3
Total number		Number for each child		Number of children

Answer : **3 children**

⚘ Summary

When the total number of cookies is distributed so that each child gets an equal number, the number of children who can share the cookies is also represented by a division sentence.

Want to confirm

1 Let's write a math sentence for each situation below and find the number of children who can share the blocks.

① Divide **9** blocks so that each child gets **3** blocks.

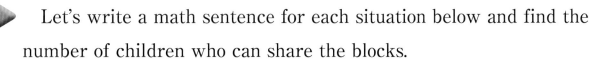

☐	÷	☐	=	☐
Total number		Number for each child		Number of children

② Divide **20** blocks so that each child gets **4** blocks.

☐	÷	☐	=	☐
Total number		Number for each child		Number of children

③ Let's try changing the number of blocks and for each child.

2 If you divide 15 candies so that each child gets 3 candies, how many children can share the candies?

① Write a math expression.

$$\boxed{} \div \boxed{}$$

Total number   Number for each child

② Let's think about how to calculate.

Ⓨ Purpose Let's think about how to find the answer of a division to find the number of children.

For 3 children   3 × 3 = 9

For 4 children   3 × 4 = 12

For 5 children   3 × 5 = 15

Number for each child   Number of children   Total number

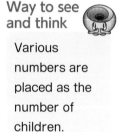

Way to see and think

Various numbers are placed as the number of children.

The answer to 15 ÷ 3 is the number to be placed in $\boxed{}$ for

3 × $\boxed{}$ = 15.

Math Sentence : 15 ÷ 3 = 5   <u>Answer : 5 children</u>

Ⓢ Summary

The answer to 15 ÷ 3 can be found by using the row of 3 in the multiplication table.

$$15 \div 3 = \boxed{}$$
$$3 \times \boxed{3} = 9$$
$$3 \times \boxed{4} = 12$$
$$3 \times \boxed{5} = \boxed{15}$$

2 There are 30 dL of milk. If you drink 6 dL a day, in how many days can you drink the milk?

[　　] ÷ [　　] = [　　]

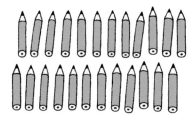

30dL

) one day

3 There are 24 pencils. If 6 pencils are placed in each box, how many boxes will be needed?

That's it. 💡 Division and multiplication

In a multiplication sentence,

Number for each child × Number of children = Total number

An example of a problem to find the total number
There are 5 children and each one is given 3 candies. How many candies are needed altogether?

If you want to find the number for each child or the number of children, division is used for finding it.

【Division to find the number for each child 】

Total number ÷ Number of children = Number for each child

An example of a problem to find the number for each child
Divide 15 candies equally among 3 children. How many candies does each one get?

【Division to find the number of children】

Total number ÷ Number for each child = Number of children

An example of a problem to find the number of children
If you divide 15 candies so that each child gets 3 candies, how many children can share the candies?

❸ 2 types of division problems

Want to compare

1 Yui and Hiroto looked at the picture on the right and made problems for 10 ÷ 5 respectively. Let's compare the differences between them by expressing it in diagrams or expressions.

Yui's problem

Problem

Divide 10 tomatoes equally into 5 plates. How many tomatoes are there on each dish?

Hiroto's problem

There are 10 tomatoes. 5 tomatoes are placed on each plate. How many plates are needed?

① Let's circle tomatoes in the diagrams below so as to represent each problem.

Diagram

② Let's find the answer to each problem.

Yui's method

Math expression

I thought about the operation to find the number ☐ in
"☐ × 5 = 10."
So, 10 ÷ 5 = ☐

Answer : ☐ tomatoes

Hiroto's method

I thought about the operation to find the number ☐ in
"5 × ☐ = 10."
So, 10 ÷ 5 = ☐

Answer : ☐ plates

The answer to a division problem can be found by using the row of the divisor in the multiplication table.

10	÷	5	=	2
Dividend		Divisor		Answer

Want to confirm

 Let's make two math problems for 24 ÷ 8.

4 Division with 1 and 0

Want to think

1 There are cookies in a box to be shared equally by 4 children.

How many cookies does each child get?

Cookies

① If there are 12 cookies, $12 \div 4 = \boxed{}$

② If there are 4 cookies, $4 \div 4 = \boxed{}$

③ If there are no cookies, $0 \div 4 = \boxed{}$

Want to confirm

 There is 6 dL juice to be poured into 1 dL cups. How many cups are needed?

2 Let's calculate the following.

① $6 \div 6$ ② $7 \div 7$ ③ $2 \div 2$ ④ $5 \div 5$

⑤ $0 \div 8$ ⑥ $0 \div 3$ ⑦ $5 \div 1$ ⑧ $9 \div 1$

Juice

6dL

Want to solve

Activity

1 There are 80 sheets of colored paper that are to be shared equally by 4 children.

How many sheets does each child get?

① Let's write a math expression.

② Nanami and Daiki think about how to calculate as shown below. Let's continue to write each idea in your notebook and explain it. Then, find the answer by using each idea.

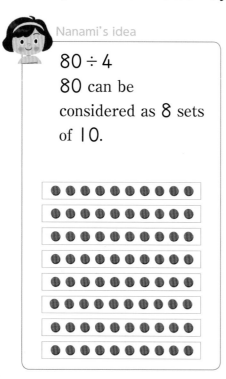

Nanami's idea

80 ÷ 4
80 can be considered as 8 sets of 10.

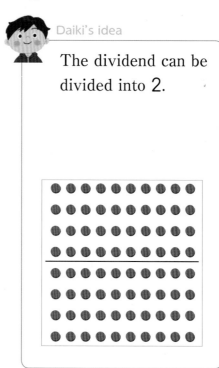

Daiki's idea

The dividend can be divided into 2.

Way to see and think

80 can be considered as 8 sets of 10.

Way to see and think

As in multiplication, a dividend can be divided in division.

Want to confirm

 Let's calculate the following.

① 90 ÷ 3   ② 40 ÷ 4   ③ 80 ÷ 2

Activity

2

Let's think about how to calculate 36 ÷ 3.

Want to explain

① Yui and Hiroto thought about how to calculate as shown below. Let's explain each idea.

Yui's idea

I thought of the row of 3 in the multiplication table.

$3 \times \boxed{9} = 27$

This is still not enough to make 36. So,

$3 \times \boxed{10} = 30$
$3 \times \boxed{11} = 33$ } + 3
$3 \times \boxed{12} = 36$ } + 3

Thus, 36 ÷ 3 = 12   Answer : 12

Way to see and think

In multiplication, if the multiplier increases by 1, the answer increases by the multiplicand.

Hiroto's idea

I thought of decomposing 36 into 30 and 6.
  30 ÷ 3 = 10
  6 ÷ 3 = 2
Then,
10 + 2 = 12
Thus, 36 ÷ 3 = 12   Answer : 12

② By using each idea, let's explain how to calculate 24 ÷ 2 and 39 ÷ 3.

Want to confirm

2   Let's calculate the following.

① 48 ÷ 4   ② 63 ÷ 3   ③ 33 ÷ 3

What you can do now

☐ Can find the answer to division by using the multiplication table.

1 Let's calculate the following.

① $27 \div 3$ ② $12 \div 2$ ③ $18 \div 2$ ④ $56 \div 7$

⑤ $40 \div 8$ ⑥ $45 \div 9$ ⑦ $63 \div 9$ ⑧ $25 \div 5$

⑨ $16 \div 4$ ⑩ $49 \div 7$ ⑪ $28 \div 7$ ⑫ $54 \div 9$

⑬ $7 \div 1$ ⑭ $9 \div 9$ ⑮ $0 \div 6$ ⑯ $2 \div 1$

⑰ $3 \div 3$ ⑱ $0 \div 9$ ⑲ $4 \div 1$ ⑳ $0 \div 1$

㉑ $70 \div 7$ ㉒ $60 \div 2$ ㉓ $88 \div 4$ ㉔ $96 \div 3$

☐ Can find the number for each.

2 Let's answer the following.

① There are 36 strawberries that are to be shared equally by 4 children. How many strawberries does each child get?

② Divide 24 sheets of origami paper equally into 3 bags. How many sheets are in each bag?

☐ Can find the number of units.

3 Let's answer the following.

① There are 24 pencils. When they are divided into 4 pencils each in bundles, how many bundles will be made?

② There is a 8 m ribbon. When it is cut into 1 m pieces, how many pieces are there?

☐ Understanding which should be found, the number for each or the number of units.

4 There is 14 dL juice. Let's think about the following.

① When each child gets 7 dL, how many children can share the juice?

② Divide the juice equally among 7 children, how many dL does each child get?

Supplementary problems p.130

Usefulness and efficiency of learning

1 Let's calculate the following.

① $35 \div 7$ ② $72 \div 8$ ③ $16 \div 8$

④ $48 \div 6$ ⑤ $12 \div 3$ ⑥ $45 \div 5$

⑦ $10 \div 2$ ⑧ $35 \div 5$ ⑨ $64 \div 8$

⑩ $36 \div 6$ ⑪ $4 \div 2$ ⑫ $16 \div 2$

⑬ $81 \div 9$ ⑭ $63 \div 7$ ⑮ $42 \div 6$

⑯ $3 \div 1$ ⑰ $8 \div 8$ ⑱ $0 \div 2$

⑲ $50 \div 5$ ⑳ $77 \div 7$ ㉑ $84 \div 2$

Can find the answer to division by using the multiplication table.

2 30 children are divided into 5 groups.

Can find the number for each.

How many children are there in each group?

3 9 cards can be made of one sheet of drawing paper.

Can find the number of units.

How many sheets of drawing paper are needed to make 36 cards?

4 Make a math problem for $32 \div 4$.

Understanding which should be found, the number for each or the number of units.

Let's fill in the ☐ with numbers or words.

Division to find the number of pencils for each child

There are ☐ pencils that are divided equally among ☐ children. How many pencils does ☐ get?

Division to find the number of children

There are ☐ pencils. Each child gets ☐ pencils. How many ☐ can share the pencils?

Is it divisible?

Put 5 apples into each bag.

25 apples

Put 6 oranges into each bag.

20 oranges

There are exactly 5 bags of apples.

25 apples

There are extra oranges.

20 oranges

Problem Let's think about how to calculate when the number cannot be divided exactly.

4 Division with Remainders
Let's think about the meaning of remainder in division.

If I put 6 oranges into each ....

How many bags can be filled?

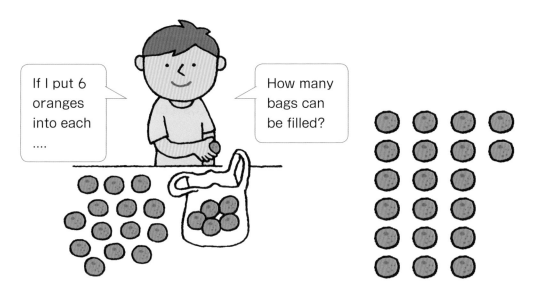

Way to see and think

Let's arrange in sets of 6.

1 Division with remainders

Want to solve

1 There are 20 oranges. If you put 6 oranges into each bag, how many bags will be filled with 6 oranges?

① Let's write a math expression.   ☐ ÷ ☐

② Look at the above picture. How many bags will be filled? How many oranges will remain?

③ Let's think about how to calculate.

Nanami: Can I find the answer by using the multiplication table?

Hiroto: Is there a multiplication sentence for "6 × ☐ = 20"?

 Let's think about how to calculate 20 ÷ 6.

By using the multiplication table, it is shown as the following.

Number of oranges in each bag		Number of bags		Total number	
1 bag	6 ×	1	=	6	14 oranges remain
2 bags	6 ×	2	=	12	8 oranges remain
3 bags	6 ×	3	=	18	2 oranges remain
4 bags	6 ×	4	=	24	4 oranges short

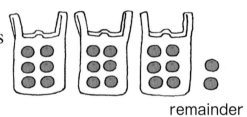

remainder

There are 20 oranges. If you put 6 oranges into each bag, there will be 3 bags and 2 remaining oranges.

This can be written as follows : 20 ÷ 6 = 3 remainder 2

Answer : 3 bags and 2 oranges remain

When there is a **remainder** after dividing, like 20 ÷ 6, the division expression is called **not divisible**. When there is no remainder, like 18 ÷ 6 and 24 ÷ 6, the division expression is called **divisible**.

Summary

Even when the dividend is not divisible, the answer to a division problem can be found by using the multiplication table.

1 ▶ Let's calculate the following.

① 11 ÷ 2   ② 48 ÷ 7   ③ 17 ÷ 3   ④ 65 ÷ 8

2 There are **34** cards. If **6** children get the same number of cards, how many cards will each child get? How many cards will remain?

Way to see and think

In division to find the number for each child, a dividend can sometimes not be divisible.

Want to think Divisor and size of the remainder

2 There are **23** chestnuts. If you put **4** chestnuts into each bag, how many bags will be filled? How many chestnuts will remain?

① Let's write a math expression.

Want to discuss

② Let's discuss the ideas of the following children.

Daiki's idea

Number in each bag	Number of bags	Total number	
4 ×	3	= 12	11 remainders
4 ×	[4]	= 16	7 remainders

23 ÷ 4 = 4 remainder 7

Answer :

4 bags and 7 chestnuts remain

Nanami's idea

Number in each bag	Number of bags	Total number	
4 ×	4	= 16	7 remainders
4 ×	[5]	= 20	3 remainders
4 ×	6	= 24	1 short

23 ÷ 4 = 5 remainder 3

Answer :

5 bags and 3 chestnuts remain

⊙Purpose In division, what relationship is there between the divisor and size of the remainder?

3 Divisions, wherein the divisor is 4, are lined up on the right. Let's fill in the ☐ with numbers and examine the relationship between the divisor and size of the remainder.

Then, let's explain what you found out about the remainders.

What does "remainder 0" mean?

Yui

Dividend	Divisor	Answer	Remainder
12 ÷	4 =	3	
11 ÷	4 =	2	Remainder 3
10 ÷	4 =	2	Remainder 2
9 ÷	4 =	2	Remainder 1
8 ÷	4 =	2	
7 ÷	4 =	1	Remainder ☐
6 ÷	4 =	1	Remainder ☐
5 ÷	4 =	1	Remainder ☐
4 ÷	4 =	1	
3 ÷	4 =	☐	Remainder ☐
2 ÷	4 =	☐	Remainder ☐
1 ÷	4 =	☐	Remainder ☐

Summary

The remainder in division should always be smaller than the divisor.

Way to see and think

When the dividend is changed to consecutive numbers, the relationship between the answer and the remainder is made clear.

4 Let's calculate the following, paying attention to the size of the remainder.

① $7 ÷ 2$ ② $10 ÷ 3$ ③ $14 ÷ 4$

④ $38 ÷ 7$ ⑤ $43 ÷ 5$ ⑥ $58 ÷ 6$

5 There are 41 chestnuts. If 5 children get the same number of chestnuts, how many chestnuts will each child get? How many chestnuts will remain?

Way to see and think

Is the remainder smaller than the divisor?

3 There are 26 candies. If you put 8 candies into each bag, how many bags will be filled? How many candies will remain?

① Let's fill in the ☐ with numbers.

☐ ÷ ☐ = ☐ remainder ☐

Total number   Number in each bag   Number of bags   Remainder

Answer : ☐ bags and ☐ candies remain

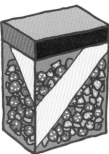

Want to think

② The answer to the above division problem can be confirmed by the following operation. Let's think about the reason.

8 × 3 + 2 = ☐

Number in each bag   Number of bags   Remainder   Total number

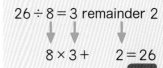
$26 ÷ 8 = 3$ remainder 2

$8 × 3 + \quad 2 = 26$

Want to confirm

6 Let's confirm the answers of the following. If the answer is not correct, write the correct answer.

① 29 ÷ 4 = 7 remainder 1   ② 19 ÷ 4 = 5 remainder 1
③ 34 ÷ 7 = 4 remainder 6   ④ 50 ÷ 6 = 7 remainder 8

 7 Let's calculate the following and confirm the answers.

① 7 ÷ 4   ② 47 ÷ 9   ③ 50 ÷ 7

Division algorithm in vertical form

> There are 25 sheets of origami paper. If each child gets 3 sheets, how many children can share the origami paper? How many sheets will remain?

① Let's write a math expression.

② Let's calculate.

As well as in addition and subtraction, we can make division algorithm in vertical form. It is shown below.

Division algorithm for 25 ÷ 3 in vertical form

$3\overline{)25}$

How to write : $3\overline{)25}$

$25 \rightarrow)25 \rightarrow \overline{)25} \rightarrow 3\overline{)25}$

divide

Write 8 above the digit in the ones place of 25.

multiply

Write 24, which is the answer to 3×8, below 25. Align the digits of the numbers according to their places.

The result when 24 is subtracted from 25 is 1.

We see that 1 is already smaller than the divisor 3.

subtract

the number of children

remainder

24 is the number of sheets given to the children.

Nanami

According to the above division algorithm in vertical form,

$25 \div 3 = 8$ remainder 1   Answer : 8 children and 1 sheet remains

Want to think

40 balls are placed in boxes. Each box has 6 balls. How many boxes do you need?

① Let's write a math expression.

② Let's think about the remainder and find the answer.   Answer : ☐ boxes

You need one more box for remaining balls.

Hiroto

Want to confirm

There is a book with 62 pages. If you read 7 pages every day, how many days do you need to finish reading?

Way to see and think

You need one more day to read remaining pages.

2 There are 28 children in Eita's class. Let's answer the following.

① If the class is divided into groups of 5 children, how many groups are formed? How many children do not belong to any group?

② If the class is divided into groups of 5 children or groups of 6 children without a remainder, how many groups of 5 and groups of 6 are formed?

Want to try

There are 29 oranges. If 9 children get the same number of oranges, what is the largest number of oranges each child can get?

Let's make a math problem for $35 ÷ 4$.

What you can do now

☐ Can solve divisions with remainder.

1 Let's calculate the following.

① $33 \div 8$ ② $48 \div 5$ ③ $17 \div 4$

④ $26 \div 7$ ⑤ $56 \div 9$ ⑥ $43 \div 6$

⑦ $13 \div 2$ ⑧ $39 \div 7$ ⑨ $74 \div 9$

☐ Can confirm the answer in a division.

2 Let's confirm the answer in the following divisions.

① $19 \div 3 = 6$ remainder 1

$3 \times \boxed{} + \boxed{} = \boxed{}$

② $31 \div 4 = 7$ remainder 3

$\boxed{} \times \boxed{} + \boxed{} = \boxed{}$

☐ Can make a division expression and find the answer.

3 There are 66 cards. Let's answer the following.

① If 9 children get the same number of cards, how many cards will each child get? How many cards will remain?

② If each child get 9 cards, how many children can share the cards? How many cards will remain?

☐ Understanding the relationship between the divisor and size of the remainder.

4 Let's find the mistakes in the following divisions.

Write the correct answers in the $\boxed{}$.

① $37 \div 5 = 8$ remainder 3

$\boxed{}$

② $28 \div 3 = 8$ remainder 4

$\boxed{}$

☐ Understanding the meaning of remainder.

5 There are 30 cakes. If 4 cakes can be placed in each box, how many boxes are needed?

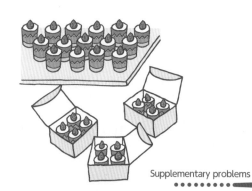

Supplementary problems p.131

Usefulness and efficiency of learning

1 Let's calculate the following and confirm the answers.

① 29 ÷ 4   ② 36 ÷ 5   ③ 17 ÷ 6

④ 36 ÷ 7   ⑤ 82 ÷ 9   ⑥ 43 ÷ 9

⑦ 55 ÷ 8   ⑧ 61 ÷ 7   ⑨ 70 ÷ 8

⑩ 7 ÷ 6   ⑪ 4 ÷ 9   ⑫ 1 ÷ 2

☐ Can solve divisions with remainder.

☐ Can confirm the answer in a division.

2 You got 46 persimmons. You would like to divide them equally among 6 people.
Let's answer the following.

① How many persimmons does each person get? How many persimmons will remain?

② How many more persimmons do you need if you want to give 8 persimmons to each person?

☐ Can make a division expression and find the answer.

3 You confirmed whether the answer to the following division problem is correct. Let's fill in the ☐ with numbers.

$$35 \div \boxed{} = \boxed{} \text{ remainder } 8 \longleftrightarrow \boxed{} \times \boxed{} + 8 = 35$$

☐ Understanding the relationship between the divisor and size of the remainder.

4 There are 29 balls to be carried. If you carry 3 balls at a time, how many times do you need to carry balls?

☐ Understanding the meaning of remainder.

5 There are 61 cookies. Let's answer the following.

① If you put 7 cookies into each bag, how many bags will be filled? How many cookies will remain?

② If you put 7 cookies or 8 cookies into each bag without a remainder, how many bags of 7 cookies and bags of 8 cookies will be filled?

☐ Can find the remainder.

Let's deepen.

What can we do by using the size of the remainder?

Hiroto

57

Develop in the 5th grade

Utilize in mathematics.

Deepen.

▲ or □ ?

Want to think

The ▲ and □ are arranged in a pattern as shown below.

▲	▲	□	□	▲	▲	□	□	▲	▲	□	□	…
1	2	3	4	5	6	7	8	9	10	11	12	

① How are the ▲ and □ arranged?

② What is the 20th shape in the arrangement?

2 pieces of ▲ and 2 pieces of □ are repeated.
Yui

Want to think

The ▲ and □ are arranged in a pattern as shown below.

□	▲	□	▲	□	□	□	▲	□	▲	□	□	…
1	2	3	4	5	6	7	8	9	10	11	12	

① What is the 20th shape?

There are 2 sets of □▲ followed by 3 pieces of □. If we continue from there, we can determine the 20th shape.
Hiroto

Maybe if we can identify the pattern, we don't have to do that.
Nanami

It looks like the pattern is □▲□▲□□.
Daiki

It's every 6 pieces, so maybe we can divide 20 by 6.
Yui

Want to deepen

Let's determine the 56th shape by using Yui's method.

That's it! The remainder in division by 9

Let's play the Remainder Game.

【How to play】

① This is played by 2 players.

② There are 36 cards with a number from 0 to 35 on each. The players draw a card each.

③ Divide the number on the card by 9 and find the remainder. Then the one who gets the larger number of the remainder wins.

Yui

My number on the card is 13.
13 ÷ 9 = 1 remainder 4

Mine is 24.
24 ÷ 9 = 2 remainder 6
So, I won.

Hiroto

The number on the card and the remainder when the number is divided by 9 are shown in the table below.

Let's fill in the blanks with the numbers and discuss what you found.

Number of the Remainder	0	1	2	3	4	5	6	7	8
Number on the Card	0	1	2	3	4	5			
	9		11		13			16	17
	18		20	21		23	24		
	27	28			31				35

$$42 \div 9 \begin{cases} 10 \div 9 = 1 \text{ remainder } 1 \\ 10 \div 9 = 1 \text{ remainder } 1 \\ 10 \div 9 = 1 \text{ remainder } 1 \\ 10 \div 9 = 1 \text{ remainder } 1 \\ 2 \div 9 = 0 \text{ remainder } 2 \end{cases}$$

remainder 4 + 2

When a number on the card is a 2-digit number, if the sum of the number in the tens place and the number in the ones place is any number from 1 to 8, the sum becomes the same as the number of the remainder when the number on the card is divided by 9.
If the sum is 9, the number of the remainder becomes 0.
If the sum is 10, it becomes 1.
If the sum is 11, it becomes 2.

 **How Many Times**

Comparing the lengths of tapes

red ███████████████████ | 2cm

blue ██████████ 6cm

yellow ▢ 3cm

Want to know

1 There are 3 tapes as shown above. Let's think about the lengths of these tapes.

① How many times of the length of the blue tape is that of the red tape?

red ████████████████

blue █████┄┄┄┄┄┄┄┄

$12 ÷ 6 =$ ▢   Answer : ▢ times

Way to see and think

I unit, 2 units and 3 units is called I time, 2 times and 3 times.

If 6 cm is regarded as I unit, I2 cm is 2 units of 6 cm.
This is called "I2 cm is **2 times** of 6 cm."

② Let's think of the blue tape as I and write the scales on the line below.

red ███████████████

blue ████████┄┄┄┄┄┄

0   (times)

60

③ How many times of the length of the yellow tape is that of the red tape?

1 The pink tape is **15 cm** long and **5** times of the green tape. How long is the green tape?

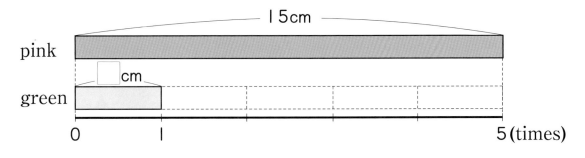

$15 ÷ 5 =$ ☐   Answer : ☐ cm

To find how many times of unit length and to find unit length, division is used.

2 Let's find **4** times of the following lengths.

Way to see and think

To find the total number, multiplication is used.

①

$2 × 4 =$ ☐

②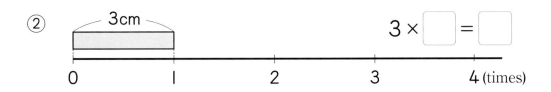

$3 ×$ ☐ $=$ ☐

How much is the total cost?

Which should I buy?

candy 15 yen
gumdrop 48 yen
jelly 24yen
gum 27yen

cookie 56 yen
cracker 215 yen
chocolate 143 yen

I bought a gumdrop and a cookie.

48 yen   56 yen

I bought a cracker and a cup of jelly.

215 yen   24 yen

The total cost was 104 yen.

The total cost was 239 yen.

I bought a cracker and a chocolate. How much was the total cost?

215 yen   143 yen

Problem Let's think about how to calculate 215 + 143.

5 Addition and Subtraction
Let's think about how to calculate 3-digit numbers in vertical form.

1 Addition of 3-digit numbers

Want to solve   Addition without carrying

Activity

1

You bought a cracker for 2 1 5 yen and a chocolate for 1 4 3 yen. How much was the total cost?

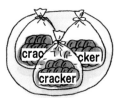

cracker
215 yen

chocolate
143 yen

Way to see and think

Remember the diagram for addition you have learned in the 2nd grade.

① Let's draw a diagram.

② Let's write a math expression.

③ Approximately, how much is it?

☐ yen

Is 400 yen enough?

Yui

Want to think

④ Let's think about how to calculate.

I'll try to use the same idea as in adding 2-digit numbers by using blocks.

Nanami

Can I add in vertical form?

Daiki

Way to see and think

Can you think of it like addition you have learned in the 2nd grade?

 Purpose   How can we add 3 digit-numbers in vertical form?

Nanami's idea

Align the blocks vertically according to their places and then add the numbers in each place.

Hundreds	Tens	Ones

boxes of 100   boxes of 10   single blocks

$2 + 1 = 3$ $1 + 4 = 5$ $5 + 3 = 8$

$215 + 143 = 358$

Daiki's idea

Use addition in vertical form like what we did in addition of 2-digit numbers.

```
    2 1 5
 +  1 4 3
    3 5 8
```

Addition algorithm for $215 + 143$ in vertical form

```
   2 1 5
 + 1 4 3
```

→

```
   2 1 5
 + 1 4 3
   3 5 8
```

Align the digits of the numbers according to their places.

$2 + 1 = 3$ $1 + 4 = 5$ $5 + 3 = 8$

Add the numbers in the same places.

Way to see and think

Also for adding large numbers, align the digits of the numbers according to their places.

Summary

For adding 3-digit numbers in vertical form, align the digits of the numbers according to their places and then add the numbers in the same places.

Want to confirm

 1

Let's calculate the following in vertical form.

① $153 + 425$ ② $261 + 637$ ③ $437 + 302$ ④ $502 + 207$

2

The number of books borrowed from the library in April was 217 and the number in May was 326. What is the total number of books borrowed in two months?

① Let's write a math expression.

② Let's think about how to add in vertical form.

```
  2 1 7        2 1 7        2 1 7        2 1 7
+ 3 2 6   →  + 3 2 6   →  + 3 2 6   →  + 3 2 6
                   ¹3          4¹3        5 4¹3
```

Don't forget to write down the number 1 that you carried.

Let's calculate the following in vertical form.

① 258 + 234   ② 512 + 249   ③ 308 + 415   ④ 102 + 418

Let's explain how to calculate 275 + 564 in vertical form.

```
  2 7 5        2 7 5        2 7 5        2 7 5
+ 5 6 4   →  + 5 6 4   →  + 5 6 4   →  + 5 6 4
                     9        ¹3 9        8¹3 9
```

Let's calculate the following in vertical form.

① 324 + 195   ② 253 + 574   ③ 625 + 190   ④ 576 + 62

3 Let's think about how to calculate $248 + 187$ in vertical form.

Addition algorithm for $248 + 187$ in vertical form

Hundreds	Tens	Ones
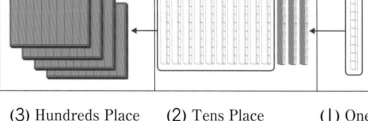		

(3) Hundreds Place
   $2 + 1 + 1 = 4$

(2) Tens Place
   $4 + 8 + 1 = 13$
   Carry 10 tens to
   the hundreds place
   as 1 hundred.

(1) Ones Place
   $8 + 7 = 15$
   Carry 10 ones
   to the tens
   place as 1 ten.

```
   2 4 8
 + 1 8 7
```

```
   2 4 8
 + 1 8 7
      ¹5
```

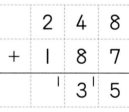

```
   2 4 8
 + 1 8 7
    ¹3¹5
```

```
   2 4 8
 + 1 8 7
   4¹3¹5
```

 Want to confirm

▶ **5** Let's calculate the following in vertical form.

① $376 + 546$ ② $468 + 254$ ③ $453 + 367$ ④ $859 + 51$

6 Let's explain how to calculate $346 + 257$ in vertical form.

```
    3 4 6          3 4 6          3 4 6          3 4 6
  + 2 5 7    →   + 2 5 7    →   + 2 5 7    →   + 2 5 7
                   ¹  3           ¹0¹ 3         6¹0¹ 3
```

7 Let's calculate the following in vertical form.

① $537 + 168$ ② $456 + 344$ ③ $737 + 68$

Addition with carrying to the thousands place

4 Let's think about how to calculate the following in vertical form.

① $856 + 327$

```
    8 5 6          8 5 6          8 5 6          8 5 6
  + 3 2 7    →   + 3 2 7    →   + 3 2 7    →   + 3 2 7
                   ¹  3            8¹ 3        1 1 8¹ 3
```

② $459 + 563$

```
    4 5 9          4 5 9          4 5 9          4 5 9
  + 5 6 3    →   + 5 6 3    →   + 5 6 3    →   + 5 6 3
                   ¹  2           ¹2¹ 2        1 0¹2¹ 2
```

8 Let's calculate the following in vertical form.

① $643 + 628$ ② $747 + 563$ ③ $526 + 474$

④ $888 + 345$ ⑤ $872 + 129$ ⑥ $906 + 95$

How much money is left?

Problem Let's think about how to calculate 328 – 215.

❷ Subtraction of 3-digit numbers

Want to solve Subtraction without borrowing

Hiroto had 328 yen. He bought a cracker for 215 yen at a sweetshop.
How much is left?

① Let's draw a diagram.

> **Way to see and think**
> Remember the diagram for subtraction you have learned in the 2nd grade.

② Let's write a math expression.

③ Approximately, how much is left?

 yen

Is more than 100 yen left?

Yui

Want to think

④ Let's think about how to calculate.

Nanami: I'll try to use the same idea as in subtracting 2-digit numbers by using blocks.

Hiroto: Can I subtract in vertical form?

> **Way to see and think**
> Can you think of it like subtraction you have learned in the 2nd grade?

ⓨ Purpose How can we subtract 3 digit-numbers in vertical form?

 Nanami's idea

Subtract the numbers in each place.

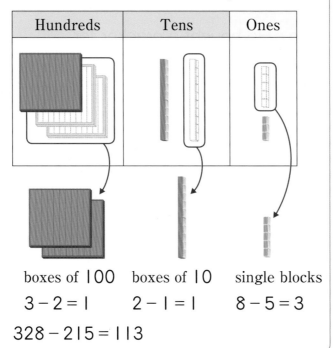

Hundreds	Tens	Ones

boxes of 100   boxes of 10   single blocks

3 − 2 = 1   2 − 1 = 1   8 − 5 = 3

328 − 215 = 113

 Hiroto's idea

Use subtraction in vertical form like what we did in subtraction of 2-digit numbers.

```
    3  2  8
 -  2  1  5
    1  1  3
```

Subtraction algorithm for 328 − 215 in vertical form

```
  3 2 8
- 2 1 5
```

→

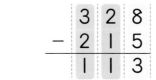
```
  3 2 8
- 2 1 5
  1 1 3
```

Align the digits of the numbers according to their places.

3 − 2 = 1   2 − 1 = 1   8 − 5 = 3
Subtract the numbers in the same places.

 Way to see and think

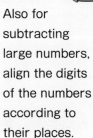

Also for subtracting large numbers, align the digits of the numbers according to their places.

Summary

For subtracting 3-digit numbers in vertical form, align the digits of the numbers according to their places and then subtract the numbers in the same places.

Want to confirm

 1 Let's calculate the following in vertical form.

① 768 − 534   ② 879 − 412   ③ 647 − 317   ④ 965 − 864

2 Let's think about how to calculate $342 - 128$ in vertical form.

```
    3 4 2           3 4̸ 2           3 4̸ 2           3 4̸ 2
  - 1 2 8    →    - 1 2 8    →    - 1 2 8    →    - 1 2 8
  ─────────        ─────────        ─────────        ─────────
                         4               1 4             2 1 4
```

(above right columns marked: 3 10)

Way to see and think

Borrow 1 ten from the tens place as 10 ones.

 2 Let's calculate the following in vertical form.

① $363 - 114$ ② $432 - 217$ ③ $540 - 513$

 3 Let's explain how to calculate $825 - 451$ in vertical form.

```
    8 2 5           8 2 5           8̸ 2 5           8̸ 2 5
  - 4 5 1    →    - 4 5 1    →    - 4 5 1    →    - 4 5 1
  ─────────        ─────────        ─────────        ─────────
                         4               7 4             3 7 4
```

(above right columns marked: 7 10)

Way to see and think

Borrow 1 hundred from the hundreds place as 10 tens.

 4 Let's calculate the following in vertical form.

① $629 - 351$ ② $257 - 183$ ③ $905 - 375$

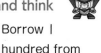

3 Let's think about how to calculate $425 - 286$ in vertical form.

Subtraction algorithm for $425 - 286$ in vertical form

Hundreds	Tens	Ones

$$\begin{array}{r} 4\,2\,5 \\ -\,2\,8\,6 \\ \hline \end{array}$$

Borrow 1 ten from the tens place as 10 ones.

① $15 - 6$

$$\begin{array}{r} {\scriptstyle 1\ 10} \\ 4\,2\,5 \\ -\,2\,8\,6 \\ \hline \square \end{array}$$

Borrow 1 hundred from the hundreds place as 10 tens.

② $11 - 8$

$$\begin{array}{r} {\scriptstyle 10} \\ {\scriptstyle 3\ \ 1\ 10} \\ 4\,2\,5 \\ -\,2\,8\,6 \\ \hline \square\,9 \end{array}$$

③ $3 - 2$

$$\begin{array}{r} {\scriptstyle 10} \\ {\scriptstyle 3\ \ 1\ 10} \\ 4\,2\,5 \\ -\,2\,8\,6 \\ \hline \square\,3\,9 \end{array}$$

5 Let's calculate the following in vertical form.

① $424 - 185$ ② $821 - 373$ ③ $510 - 176$

④ $420 - 235$ ⑤ $242 - 64$ ⑥ $740 - 69$

4 Let's think about how to calculate 305 − 178 in vertical form.

Hundreds	Tens	Ones

$$\begin{array}{r} 3\ 0\ 5 \\ -\ 1\ 7\ 8 \\ \hline \end{array}$$

Borrow 1 hundred from the hundreds place as 10 tens.

Borrow 1 ten from the tens place as 10 ones.

① 15 − 8

$$\begin{array}{r} {\scriptstyle 9} \\ {\scriptstyle 2\ \not10\ 10} \\ 3\ 0\ 5 \\ -\ 1\ 7\ 8 \\ \hline \bigcirc \end{array}$$

③ 2 − 1   ② 9 − 7

$$\begin{array}{r} {\scriptstyle 9} \\ {\scriptstyle 2\ \not10\ 10} \\ 3\ 0\ 5 \\ -\ 1\ 7\ 8 \\ \hline \bigcirc\ \bigcirc\ 7 \end{array}$$

6 Let's think about how to calculate 1000 − 895.

	1	0	0	0
−		8	9	5

7 Let's calculate the following in vertical form.

① 405 − 286   ② 402 − 107   ③ 800 − 197

④ 1000 − 536   ⑤ 1041 − 784   ⑥ 1237 − 414

Want to know

1
Let's think about how to find the answer for larger numbers by using what you have already learned.

① 4175 + 3658

	4	1	7	5
+	3	6	5	8

② 6073 + 2981

	6	0	7	3
+	2	9	8	1

③ 7008 + 2992

	7	0	0	8
+	2	9	9	2

④ 3925 − 1947

	3	9	2	5
−	1	9	4	7

⑤ 3007 − 2639

	3	0	0	7
−	2	6	3	9

⑥ 10000 − 5089

	1	0	0	0	0
−		5	0	8	9

Even if we deal with larger numbers, whenever we start adding and subtracting the numbers from the ones place, we can get the answer.

Want to confirm

 Let's calculate the following in vertical form.

① 4563 + 3125　② 2606 + 3198　③ 3587 + 6413

④ 6497 − 2135　⑤ 8114 − 3518　⑥ 10000 − 6001

Activity

1 Let's calculate the following in easier ways.

Ⓐ $298 + 120$ Ⓑ $500 - 198$

You can calculate without vertical form.

Hiroto

298 is approximately 300.

198 is approximately 200.

Nanami

① Let's explain the ideas of the following children.

Hiroto's idea

I think
$298 + 120 = 300 + 118$.

Nanami's idea

I think
$500 - 198 = 502 - 200$.

In addition, the answer does not change by adding a number to the augend and subtracting that same number from the addend.

$298 + 120$
add 2 ↓   ↓ subtract 2
$300 + 118$

In subtraction, the answer does not change by adding the same number to both the subtrahend and the minuend.

$500 - 198$
add 2 ↓   ↓ add 2
$502 - 200$

Let's calculate the following in easier ways.

① $308 + 197$ ② $499 + 350$ ③ $199 + 299$

④ $301 - 99$ ⑤ $600 - 297$ ⑥ $200 - 95$

2 Let's think about how to calculate $875 + 47 + 53$.

Nanami: $875 + 47 = 922$
Then $922 + 53$ ....

Hiroto: $47 + 53 = 100$
So, calculate this first,
and then do $875 + 100$.

When adding 3 numbers, changing the order of addition gives the same answer.
$875 + 47 + 53 = 875 + (47 + 53)$

If you change the order of calculations, it can be easier.

2 Let's calculate the following in easier ways.

① $492 + 84 + 16$ ② $52 + 365 + 48$

3 Daiki and Yui calculated mentally.

Let's explain each idea.

① $35 + 46$ ② $81 - 27$

Daiki: Add from the higher place.
① $30 + 40 = 70$
② $5 + 6 = 11$
③ $70 + 11 = 81$

Yui: Subtract from the higher place.
① $81 - 20 = 61$
② $61 - 7 = 54$

4 Let's calculate mentally.

① $18 + 6$ ② $68 + 29$ ③ $23 - 8$ ④ $71 - 46$

Want to think

1 There are 245 red roses and 138 white roses that bloomed. Let's answer the following.

① What is the total number of roses that bloomed?

[ ] roses altogether

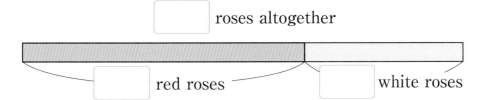

[ ] red roses   [ ] white roses

② Which color of roses is more?

[ ] red roses

difference

[ ]   [ ] roses

Want to try

▶ **1** There were 605 children in Sakura's school. During a sports day, the children were divided into the red team and the white team. There were 298 children in the red team. How many children were there in the white team?

[ ] children altogether

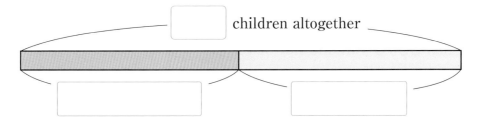

[ ]   [ ]

▶ **2** The 3rd graders collected 118 milk cartons. The 4th graders collected 20 more cartons than the 3rd graders. How many cartons did the 4th graders collect?

Let's express it in a diagram.

What you can do now

□ Can add in vertical form.

1 Let's calculate the following in vertical form.

① 324 + 253   ② 146 + 537   ③ 473 + 261

④ 246 + 485   ⑤ 354 + 249   ⑥ 464 + 368

⑦ 734 + 862   ⑧ 947 + 587   ⑨ 457 + 546

⑩ 4137 + 1425   ⑪ 2056 + 3794   ⑫ 2361 + 7639

□ Can subtract in vertical form.

2 Let's calculate the following in vertical form.

① 658 − 325   ② 374 − 138   ③ 546 − 369

④ 432 − 136   ⑤ 604 − 247   ⑥ 700 − 463

⑦ 1529 − 716   ⑧ 1153 − 645   ⑨ 1000 − 437

⑩ 3947 − 1925   ⑪ 3142 − 1734   ⑫ 10000 − 4005

□ Can calculate in easier ways.

3 Let's calculate the following in easier ways.

① 397 + 240   ② 800 − 198

③ 5387 + 57 + 43   ④ 26 + 3285 + 74

□ Can make a math expression and find the answer.

4 Ayumi has 3596 yen and her sister has 4487 yen in their savings.

① Who has more savings and by how much?

② How much are their total savings?

Supplementary problems p.133

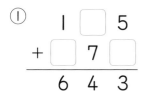

Usefulness and efficiency of learning

1 Let's fill in the ☐ with numbers.

①
```
  1 ☐ 5
+ ☐ 7 ☐
───────
  6 4 3
```

②
```
  3 ☐ 9
+ ☐ 3 ☐
───────
  6 0 0
```

③
```
  5 ☐ ☐
+ ☐ 9 6
───────
  7 0 4
```

④
```
  8 5 ☐
- 2 ☐ 5
───────
  ☐ 8 5
```

⑤
```
  ☐ ☐ 3
-   4 ☐
───────
    6 6
```

⑥
```
  9 ☐ ☐
- ☐ 5 3
───────
    8 7
```

☐ Can add in vertical form.

☐ Can subtract in vertical form.

2 Let's find the mistakes in the following calculations and write the correct answers.

①
```
  2 8 7
+ 1 4 9
───────
  4 2 6
```
(   )

②
```
  4 0 3
- 2 4 6
───────
  1 6 7
```
(   )

☐ Can add in vertical form.

☐ Can subtract in vertical form.

3 Let's calculate the following in easier ways. Also explain in what ways you did.

① 199 + 397   ② 600 − 498   ③ 351 + 97 + 49

☐ Can calculate in easier ways.

4 There are 2368 boys and 2356 girls in the elementary school at the city where Asahi lives.

How many elementary school children are there in that city altogether? Which is more and by how many?

☐ Can make a math expression and find the answer.

Let's deepen.

Can we utilize calculations of larger numbers in life?

Daiki

Deepen.

Let's go shopping!

Riku went to some stores to shop.

He brought money and a shopping list.

Shopping List:

fruits for dessert,
fruits for ingredients
of cake,
strawberry, notebook,
pencil, light bulb

Want to think

Riku went to a grocery store.

oranges   420 yen

bananas   250 yen

melon   480 yen

strawberries   390 yen

pineapple   460 yen

apple   130 yen

① He bought oranges and bananas for dessert. How much was the total cost?

② The ingredients for the cake were strawberries and others. The budget for them was at most 1000 yen. If Riku wanted to spend exactly 1000 yen, which fruits should he buy?

Next, Riku went to a stationary shop. The amount of money left with him is shown below.

① Riku bought two 60 yen pencils and a 200 yen notebook. How much was the total cost?

② If he gave a 500-yen coin, how much was his change?

③ With the amount of money left in his wallet, how should he have paid to get the least total number of coins?

Riku's last stop was at an electrical shop.

① Riku bought a 630 yen light bulb. If he gave 1000 yen, how much was his change?

② With the amount of money left in his wallet, how should he have paid to get the least total number of coins?

When Riku paid to get the least total number of coins, how many coins are left in his wallet?

 (   ) coins   (   ) coins

 (   ) coins   (   ) coins

Reflect Connect

Problem

Let's calculate in vertical form.

Let's confirm in the ways we have learned.

① 499 + 498

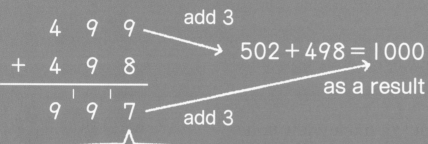

```
    4   9   9
+   4   9   8
─────────────
    9   9   7
```

add 3

502 + 498 = 1000
as a result

add 3

(3 smaller than 1000)

① Add the numbers in each place
② Starting from the ones place
③ Carry 1 to the next place

- -

② 900 − 365

(100 smaller than 1000)

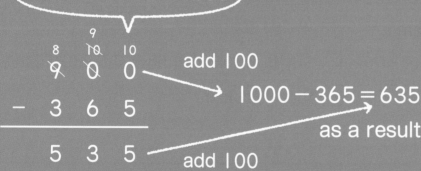

```
    8  10  10
    9   0   0
−   3   6   5
─────────────
    5   3   5
```

add 100

1000 − 365 = 635
as a result

add 100

① Subtract the numbers in each place
② Starting from the ones place
③ When we cannot subtract, borrow 1 from a higher place

Can I calculate 4-digit numbers in vertical form?

Nanami

Remember what you have learned to calculate in vertical form.
Let's try to do in the same way.

```
    5  0  2
 +  4  9  8
 ──────────
 1  0  0  0
```

I could in the same way.

- - - - - - - - - - - - - - - -

```
      9  9
     10 10 10
  1  0  0  0
 −     3  6  5
 ──────────────
     6  3  5
```

I could in the same way.

Summary

· We can do addition that has an answer larger than 1000 and subtract from 1000 in the same way as that we have learned.

· I think we can calculate numbers larger than 1000 in the same way.

We can calculate 4-digit numbers in the same way.

Hiroto

I want to try calculating various other numbers.

Yui

What kinds of vehicles passed frequently?

Many kinds of vehicles are passing.

What kinds of vehicles are passing frequently?

Let's count!

The vehicles that passed from 9:00 to 9:05 a.m. are shown at the bottom from this page to the next page.

The vehicles that passed from 9:05 to 9:10 a.m. are found at the bottom from page 87 to page 117.

Problem Let's think about how to organize and represent the data for easy understanding.

6 Tables and Graphs
Let's summarize investigations for easy understanding.

❶ Tables

Want to know

1 The table on the right shows the data about the vehicles that passed in front of the school from 9:00 to 9:05 a.m.

① Let's change the character "正" to numbers.

② Let's discuss how to make a table.

Vehicles that passed in front of the school
(9:00 to 9:05 a.m.)

Kind	Number of vehicles	
Truck	正 一	
Car	正 丁	
Bus	一	
Ambulance	一	
Total		

Want to represent

1 Let's investigate the vehicles that passed in front of the school from 9:00 to 9:10 a.m. The table below shows the data in **1**. Vehicles like an ambulance that passed infrequently are included in "others."

① Let's look at the pictures found at the bottom from page 87 to page 117 to know the vehicles that passed from 9:05 to 9:10 a.m. Add the data obtained by using the table on the right.

② How many vehicles passed from 9:00 to 9:10? Where should the number be written in the table?

③ What can we learn from this table?

Vehicles that passed in front of the school
(9:00 to 9:10 a.m.)

Kind	Number of vehicles	
Truck	正 一	
Car	正 丁	
Bus	一	
Others	一	
Total		

❷ Bar graph

1

The bar graph on the right shows the results of the investigation that Hinano and her friends did about the numbers of the vehicles that passed in front of their school.

① What does the length of bars represent?
② How many vehicles are represented by 1 cell on the graph?
③ How many trucks were there?
④ What kind of vehicles passed most frequently? How many were there?

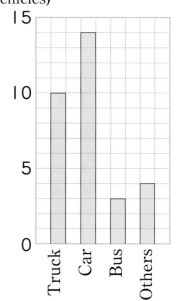

Vehicles that passed in front of the school
(9:00 to 9:10 a.m.)
(vehicles)

A graph, which uses bars of different lengths to represent data, is called a **bar graph**.

In a bar graph, we can compare the size of numbers easily.

Daiki

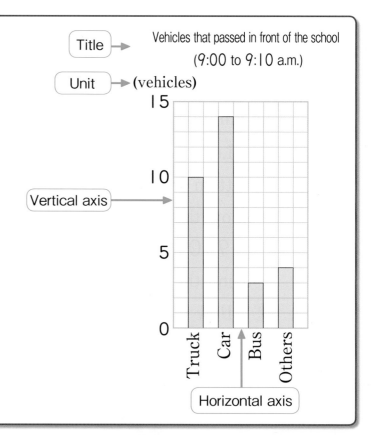

Title → Vehicles that passed in front of the school
(9:00 to 9:10 a.m.)
Unit → (vehicles)

Vertical axis

Horizontal axis

⑤ The bar graph in the previous page was changed into the one on the right. What has changed in the graph?

In the bar graph, the bars are often drawn in descending order.
The "others" bar is usually drawn last.

Vehicles that passed in front of the school (9:00 to 9:10 a.m.)

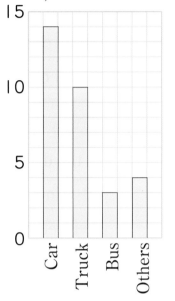

Want to compare

2 The bar graph on the right shows the results of the investigation that Kaito and his friends did about the numbers of the vehicles that passed in front of the station.

① What kind of vehicles passed the most frequently? What kind passed the least?

② What is the difference in number between the most and the least?

③ What is the total number of vehicles that passed in front of the station?

④ What can you say about this graph as compared to the graph above?

Vehicles that passed in front of the station (9:00 to 9:10 a.m.)

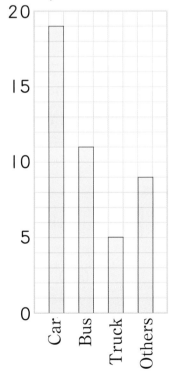

3 Sayuri and her friends investigated about the number of children who visited the nurse's office because of illness or injury. For one week in April, they recorded the number of children who visited each weekday and presented the data in a bar graph.

① In what order does the graph represent tha data?

② How many children are represented by 1 cell on the graph?

③ Let's find the number of children who visited the nurse's office each day.

④ Let's compare the numbers of children who visited the nurse's office on Monday and Thursday.

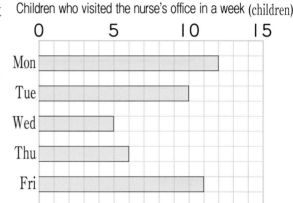

Children who visited the nurse's office in a week (children)

A bar graph is sometimes drawn horizontally.

 When the data to be presented in a bar graph are in order, like the days of the week, the bars should be drawn in that order.

1 In the graphs below, let's find the number that represents 1 cell.

① (cm)
60
40
20
0

② (L)
30
20
10
0

③ (yen)
500
0

④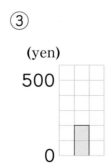
0   10 (pieces)

4 The table on the right shows the favorite sports of the 3rd grade children in class 1. Let's draw a bar graph.

Favorite sports

Sport	Number of children
Soccer	14
Baseball	10
Dodgeball	7
Swimming	3
Others	2
Total	36

How to draw a bar graph

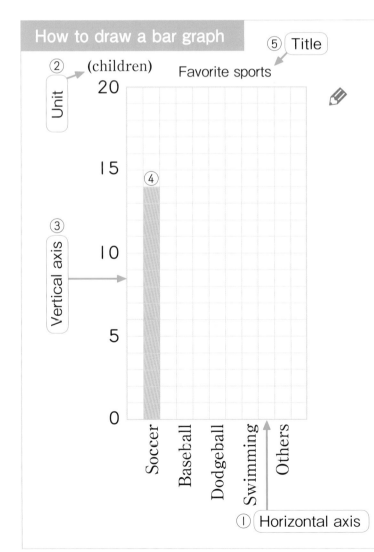

① Write each sport on the horizontal axis.

② Write the unit on the vertical axis.

③ Write the scale on the vertical axis. When deciding on the scale to use, consider the highest number of children, and write the units like 5, 10, etc.

④ Draw each bar according to the number of children.

⑤ Write the title of the graph.

89

2 The following table shows the number of the 3rd grade children in each class whose favorite sport is soccer.

Let's draw a bar graph.

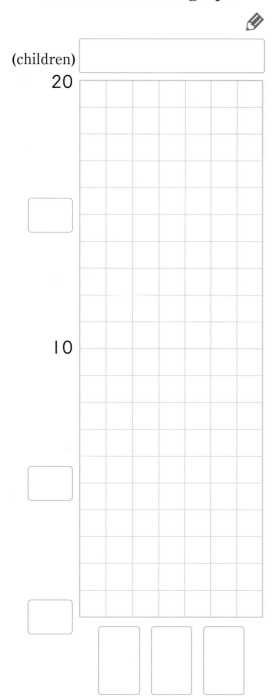

3rd grade children who like soccer

Class	Number of children
1	14
2	15
3	11
Total	40

3 The following table shows the favorite sport of all the 3rd grade children.

Let's draw a bar graph.

Favorite sports

Sport	Number of children
Soccer	40
Baseball	35
Dodgeball	15
Swimming	10
Others	5
Total	105

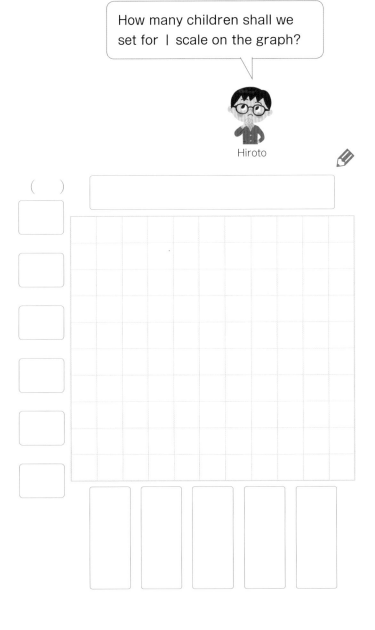

How many children shall we set for 1 scale on the graph?

Hiroto

1 The following tables show the number of the 3rd grade children who borrowed each kind of books in April, May, and June.

Books borrowed in April

Kind	Number of books
Fiction	15
Biography	6
Picture	8
Others	5
Total	

Books borrowed in May

Kind	Number of books
Fiction	21
Biography	19
Picture	24
Others	8
Total	

Books borrowed in June

Kind	Number of books
Fiction	16
Biography	14
Picture	19
Others	9
Total	

① What is the total number of books that were borrowed in each month?

② Which kind of books was borrowed the most in each month?

③ Let's organize the tables for each month into one table.

Books borrowed by the 3rd grade children   (books)

Kind \ Month	April	May	June	Total
Fiction	15	21	16	52
Biography	6	19		ⓓ
Picture	8			ⓔ
Others	5			ⓕ
Total	ⓐ	ⓑ	ⓒ	ⓖ

All we did was to put the 3 tables together.

Nanami

④ How many fiction books were borrowed from April to June?

⑤ What numbers are in boxes ⓐ, ⓑ, ⓒ, ⓓ, ⓔ, and ⓕ?

⑥ What does the number in ⓖ represent?

⑦ What kind of book was borrowed the most from April to June?

1 ▶ The table below shows the number of children in Yuma's school who got injured and the type of injury they had in April, May, and June.

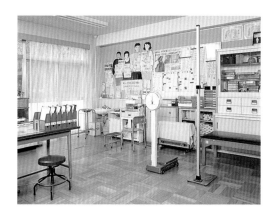

① How many children were injured in each month?

② What type of injury happened the most from April to June?

③ What does the number 46 in the table represent?

Record of injuries of children (children)

Month／Kind	April	May	June	Total
Scratch	29	27	13	
Bruise	21	46	30	
Cut	13	7	4	
Sprain	7	4	2	
Others	10	14	6	
Total				

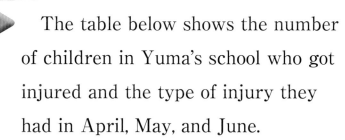

What you can do now

☐ Can read the numbers from tables or graphs and draw graphs.

1 Children collected empty cans at Koharu's school. The following table and bar graph show the number of cans collected by the children in each grade.

Empty cans collected by children

Grade	1	2	3	4	5	6	Total
Number of cans		120		240	160		

① How many cans does 1 cell represent on the graph?

② Let's write the numbers that apply in the above table.

③ Let's draw the bars for the 2nd, 4th, and 5th grades on the graph.

④ Between the table and the graph, which makes it easier to see what grade children collected empty cans the most?

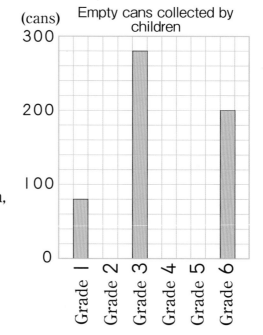

(cans) Empty cans collected by children

☐ Can read the numbers from tables.

2 The table on the right shows the number of children in each grade at Ren's school who got injured and the type of injury they had in June.

Record of injuries of children (June)   (children)

Kind \ Grade	1	2	3	4	5	6	Total
Scratch	3	ⓑ	2	5	3	4	21
Cut	ⓐ	2	2	3	ⓔ	3	ⓖ
Bruise	1	1	ⓒ	2	2	ⓕ	13
Others	2	3	1	1	0	2	9
Total	7	10	8	ⓓ	9	13	ⓗ

① Write the numbers that apply in boxes ⓐ, ⓑ, ⓒ, ⓓ, ⓔ, ⓕ, ⓖ, and ⓗ.

② What type of injury happened the most in June?

③ In what grade did the children get injured the least in June?

④ What does the number written in the box ⓗ represent?

Supplementary problems
p.134

Usefulness and efficiency of learning

1 The following table shows favorite fruits of children in Miyu's class. Let's draw a bar graph.

Favorite fruits

Fruit	Number of children
Orange	5
Apple	12
Strawberry	8
Banana	4
Others	7
Total	36

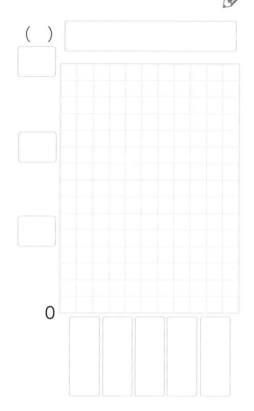

( )

0

2 The following table shows the number of the 3rd grade children in each class who borrowed each kind of books.

Books borrowed by the 3rd grade children   (books)

Kind＼Class	1	2	3	Total
Fiction	8	10	6	24
Picture	7	ⓐ	10	28
Biography	ⓑ	9	8	ⓒ
Others	3	5	ⓓ	15
Total	24	35	31	90

① Write the numbers that apply in boxes ⓐ, ⓑ, ⓒ, and ⓓ.

② What does the number 9 in the table represent?

③ Find the number that is the same as the one of the picture books borrowed by children in the class 3. What kind of book is it, and in what class?

Let's deepen.

What is important for drawing a bar graph?

Daiki

Deepen.

Proper graph?

01301

Hiroto investigated the number of typhoons which formed from 2012 to 2017 and presented the data in a bar graph.

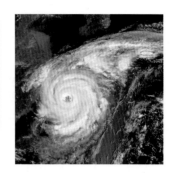

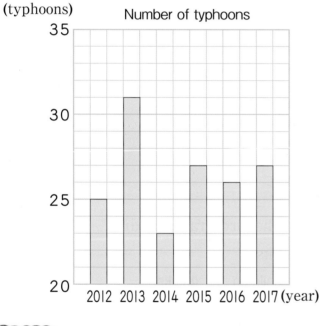

(typhoons)

Number of typhoons

Yui

This graph shows there were so many typhoons in 2013.

Daiki

The number of typhoons in 2013 is twice as many as in 2012, isn't it?

Nanami

In this graph, the units on the vertical axis don't start from 0 but from 20. Is it right?

① Three children are talking about the graph. Let's discuss how to think of their opinions.

② Let's draw the proper graph in the next page based on the table shown on the right.

Number of typhoons

Year	Number of typhoons
2012	25
2013	31
2014	23
2015	27
2016	26
2017	27

(typhoons)

Number of typhoons

2012 2013 2014 2015 2016 2017 (year)

Want to deepen

Let's compare Hiroto's graph with the graph that you drew above, and discuss what you found.

What way of thinking is used?

Four children found the number of marbles in A, B, and C. As they investigated the ways to find the number, the following math sentences were found.

A 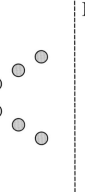   B 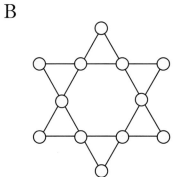   C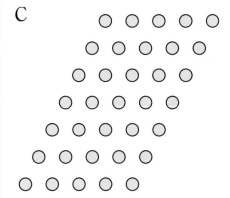

A	B	C
① 3 × 6 = 18 18 + 1 = 19 ② 7 × 3 = 21 21 − 2 = 19	① 6 × 2 = 12 ② 3 × 6 = 18 18 − 6 = 12	① 5 × 7 = 35 ② 7 × 5 = 35

How about circling marbles?

Nanami

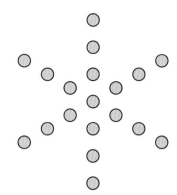

Think by myself

1 By what way of thinking did they make math sentences ① and ② for A? Let's think about it by circling marbles in the diagram on the right.

Think in groups

2 Let's think about by what way of thinking they made math sentences ① and ② either for B or for C?

> Make two groups. Let's think about B in one group and C in the other.
> If you find one way of thinking, try to find the other.

Think as a class

3 Let's make a presentation about each way for A, B, and C, and think about similarities among them.

4 Look at the marbles as shown on the right. The way to find the number of these marbles is represented as follows. Let's make a math expression for this way.

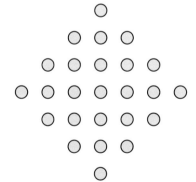

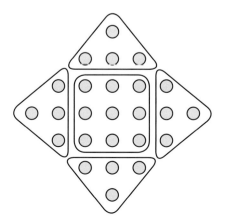

01302

How far did a rubber band car run?

Problem Let's think about how to measure a length between 2 places longer than 1 m.

Length

7 Let's think about the units of longer length and how to represent it.

 ❶ How to measure

Want to think

1 Let's think about how to measure a length that a rubber band car ran.

Nanami: It can be measured by using a meterstick about 4 times.

Yui: Can we measure in a straight line?

⚐ **Purpose** How can we measure a length longer than 1 m?

It is difficult to measure by using metersticks.

The length between two places along a straight line is called **distance**.

Tape measures are good tools for measuring the distance.

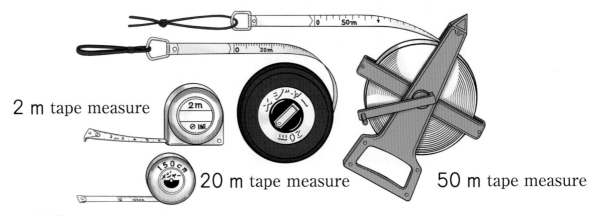

2 m tape measure

20 m tape measure

50 m tape measure

150 cm tape measure

Be careful when you adjust a 0 point of the measure to one end.

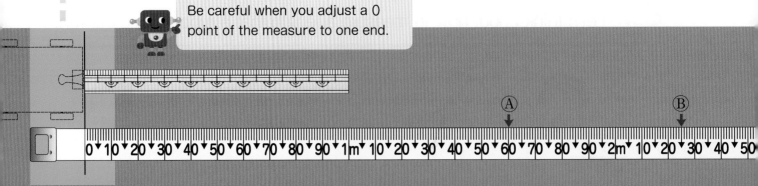

1 Let's investigate how to use a tape measure.

① Look for the location of 0 m on the tape.

Some tape measures have a 0 point, others don't.

② Let's find the distance that Yuki's car ran by reading her record shown in the tape measure below.

③ How many **m** and **cm** are the lengths of Ⓐ, Ⓑ, and Ⓒ shown on the tape measure below. And let's point ↓ at each scale that represents the length Ⓓ, Ⓔ, and Ⓕ.

Ⓓ 3m5cm Ⓔ 75cm Ⓕ 4m54cm

Summary

When we measure a distance longer than 1 m, we can use a tape measure.

 2 About how long is 10 m? Walk to a point that you think is 10 m away. Then, let's measure the actual length.

About how many steps is it?

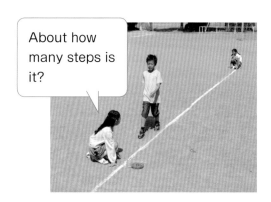

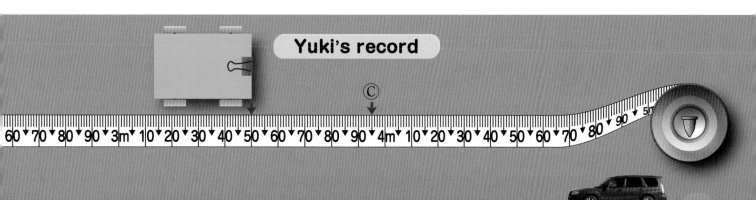

Yuki's record

Ⓒ

60 70 80 90 3m 10 20 30 40 50 60 70 80 90 4m 10 20 30 40 50 60 70 80 90 5

2 For the rulers and tape measures in the ☐ below, which is appropriate to measure the following length?

First, estimate each length.

Ⓐ The length and width of a desk
Ⓑ The length around a can
Ⓒ The length of a hallway

30 cm ruler │ m ruler (meterstick) │50 cm tape measure 50 m tape measure

3 How many cm is the length around the tree on the picture on the right?

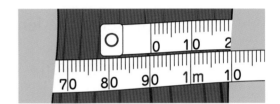

4 Let's estimate the lengths of things that can be found in our surroundings, and actually measure them.

Lengths of various things

Things measured	Estimated length	Actual length
Width of a bulletin board		
Length around a tree		
Height of an iron bar		

What tree has the longest circumference in our school?

Daiki

Write down what you found out and understood after the investigation.

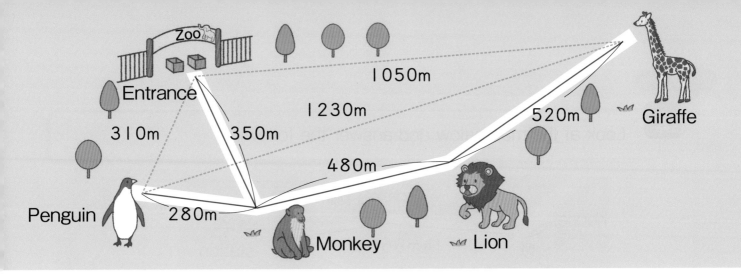

2 Kilometer

Want to think

1 Look at the above map of the zoo and answer the following.

① How many meters is the distance from the entrance to the penguins? And how many meters is the length measured along the road?

The length measured along the road is called **road distance**.

road distance

distance

② How many meters is the road distance from the monkeys to the giraffes?

1000 m is written as 1 km and is called **one kilometer**.

1km = 1000m

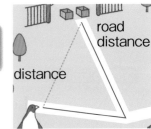

③ How many km and m are the road distance and the distance from the entrance to the giraffes respectively?

road distance   1350m = ☐ km ☐ m

distance   1050m = ☐ km ☐ m

km			m
1	3	5	0
1	0	5	0

1 km 350 m is said as "1 kilometer and 350 meters."

④ How many km and m are the road distance and the distance from the penguins to the giraffes respectively?

2 Look at the map below and answer the following.

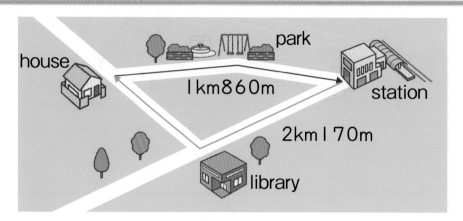

① How many **km** and **m** is the total road distance from the house to the station through the front of the park and returning to the house through the front of the library? Write a math expression.

② Let's explain the ideas of Nanami and Hiroto.

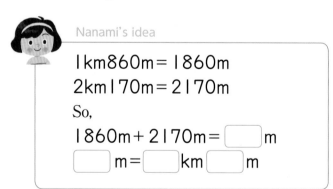

Nanami's idea

1km860m＝1860m
2km170m＝2170m
So,
1860m＋2170m＝ ☐ m
☐ m＝ ☐ km ☐ m

Hiroto's idea

	km			m
	1	8	6	0
＋	2	1	7	0

③ Which road distance from the house to the station is longer and by how much?

	km			m
	2	1	7	0
－	1	8	6	0

Where is I km from school?

3 Let's actually walk I km around the school.

① Where do you think you will end up if you walk Ikm from school?

② Let's estimate how long it takes you by walking.

③ Let's summarize your observations.

Let's walk Ikm   September 6
① Up to where did we walk?
                        Post office
② Spent Time :
                        20 minutes
③ My reflection :
    I km is longer than I expected.

How many meters are there in I km?

Yui

How many minutes will it take to walk I km?

Nanami

1 Let's look for the units of length that are used in our surroundings.

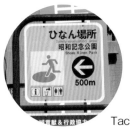

Tachikawa City, Tokyo Metropolis

What you can do now

Understanding longer lengths.

1 Let's fill in the ☐ with numbers or words.

① Choose two places and measure the length between the two places along a straight line. This is called ☐ .

② The distance measured along the road is the ☐ .

Can read length correctly by using a tape measure.

2 How many m and cm are the lengths of points, ①, ②, and ③ located on the tape measure below?

Understanding the relationship between km and m.

3 Let's fill in the ☐ with numbers.

① 1 km = ☐ m

② 2 km 50 m = ☐ m

③ 1116 m = ☐ km ☐ m

Understanding the difference between road distance and distance.

4 Look at the map below and answer the following.

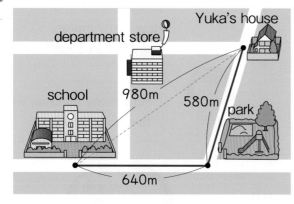

① How many km and m is the road distance from Yuka's house to the school through the front of the park?

② What is the difference between the road distance in ① and the distance from Yuka's house to the school in m?

Supplementary problems •••••••• p.136

Usefulness and efficiency of learning

1 Let's fill in the ⬜ with appropriate units.

☐ Understanding longer lengths.

① The length of a classroom from the front to the back is 8 ⬜.

② The road distance walked in 1 hour is 4 ⬜.

2 How many **m** and **cm** are the lengths of points, ⓐ~ⓕ, located on the tape measures below?

☐ Can read length correctly by using a tape measure.

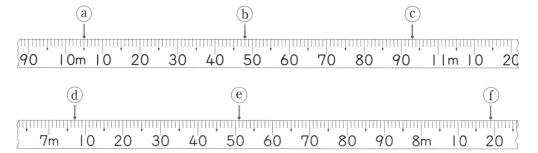

3 Let's arrange from the longest to the shortest.

☐ Understanding the relationship between km and m.

① 7 km, 700 m, 7007 m, 7 km 700 m

② 1 km 80 m, 1800 m, 1 km 8 m, 810 m

4 When Sho goes to school, he can pass through the front of either Riho's house or Akari's house. Which way has a longer road distance and by how much?

☐ Can calculate lengths.

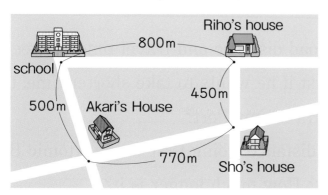

Let's **deepen.**

We understood the difference between road distance and distance.
Can we think about them relating to time?

Nanami

Deepen.

Traveling by tram

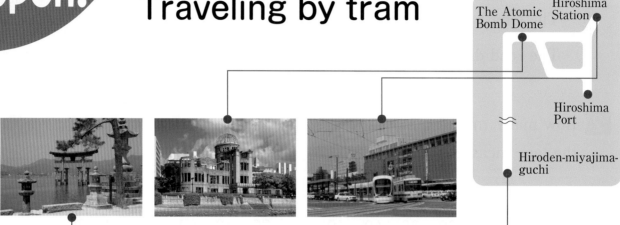

Want to think

Takuma takes a tour of Hiroshima City by tram. He leaves Hiroshima Station, visits both the Atomic Bomb Dome and the Hiroshima Port, and finally arrives at Miyajima Island.

Road distance and time

	Road distance	Time
Hiroshima Station↔Atomic Bomb Dome	2 km 400 m	17 min
Hiroshima Station↔Hiroshima Port	6 km 100 m	32 min
Atomic Bomb Dome↔Hiroshima Port	6 km 200 m	36 min
Atomic Bomb Dome↔Hiroden-miyajima-guchi	19 km 100 m	51 min

① The above table shows the road distance and travel time between two places. Where should he go first if he wants to take shorter time, to the Atomic Bomb Dome or to the Hiroshima Port?

② Which will be a longer road distance, going first to the Atomic Bomb Dome or to the Hiroshima Port? How much longer is it?

③ In ②, which takes longer by tram and by how much?

Which set is a better deal?

1000 yen for each set

With the same price set for each type of boxes, the more pieces one set has, the better deal that is.

Chocolate   Chocolate   Chocolate

6 boxes of 8 pieces   5 boxes of 10 pieces   4 boxes of 12 pieces

For a set of 8 pieces, 8 × 6 ....
For a set of 10 pieces, 10 × 5 ....

For a set of 12 pieces, there are 4 boxes that contain 12 pieces ....

Problem Let's think about how to find the answer to a multiplication which isn't found in the multiplication table.

How to Multiply (2-digit number) × (1-digit number)
Let's think about how to calculate easily.

1 There are 12 chocolates in each box. There are 4 boxes. How many chocolates are there altogether?

① Let's write a math expression.

□ × □

number of chocolates in each box   number of boxes

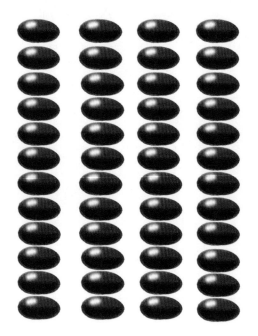

Way to see and think

A multiplication sentence is expressed as (number in each box) × (number of box) = (total number).

② Approximately, how many is the answer?

□ chocolates

③ Let's think about how to calculate 12 × 4.

The answer will be larger than that to 10 × 4.

Daiki

12 × 4 isn't found in the multiplication table.

Hiroto

It's hard to add 12 + 12 + 12 + 12.

Nanami

④ Let's explain the ideas of the following three children.

Way to see and think

You get the same answer even if you decompose the multiplicand or the multiplier.

Hiroto's idea

$$12 \times 4 \begin{cases} 6 \times 4 = 24 \\ 6 \times 4 = 24 \end{cases}$$

Total ☐

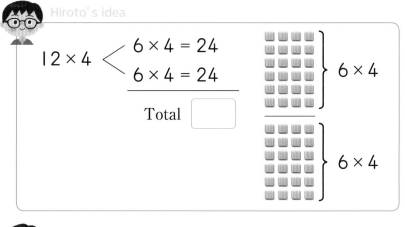

6 × 4

6 × 4

Hiroto decomposed 12 into halves. Some numbers cannot be decomposed into halves.

Daiki

Daiki's idea

$$12 \times 4 \begin{cases} 9 \times 4 = 36 \\ 3 \times 4 = 12 \end{cases}$$

Total ☐

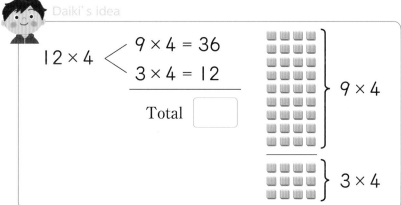

9 × 4

3 × 4

Nanami decomposed the multiplicand into the tens place and the ones place.

Yui

Nanami's idea

$$12 \times 4 \begin{cases} 2 \times 4 = 8 \\ 10 \times 4 = 40 \end{cases}$$

Total ☐

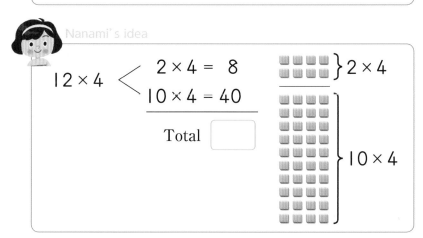

2 × 4

10 × 4

Want to connect

⑤ Let's find the answer to 18 × 4 by using various methods.

I want to try to calculate much larger numbers.

Hiroto

Multiplication with 1-digit Number
Let's think about how to multiply in vertical form.

Chocolate
20 yen apiece

Cookies
200 yen a bag

1 Multiplication with tens and hundreds

Want to know

Activity

1 Let's find the cost of the following chocolates and cookies.

Ⓐ   4 chocolates that cost 20 yen each

Ⓑ   4 bags of cookies that cost 200 yen each

① Let's write each math expression.

Ⓐ [   ] × [   ]

Cost of one chocolate   Number of chocolates

Ⓑ [   ] × [   ]

Cost of one bag   Number of bags

Want to think

② Let's think about each method for calculating.

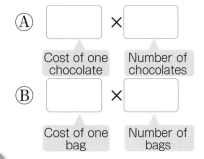

20 yen is two 10-yen coins ....

Yui

200 yen is two 100-yen coins ....

Daiki

Way to see and think

It is easier for thinking to change the costs to 10-yen coins and 100-yen coins.

 **Purpose** How can we multiply tens and hundreds?

 Yui's idea

A 20 × 4

20 is 2 sets of 10

20 × 4 is

(2 × 4) = 8 sets of 10

20 × 4 = 80

 Daiki's idea

B 200 × 4

200 is 2 sets of 100

200 × 4 is

(2 × 4) = 8 sets of 100

200 × 4 = 800

③ What are the similarities between Yui's idea and Daiki's?

They think by using sets of 10 or 100.

Hiroto

Each of them uses the multiplication 2 × 4.

Nanami

Summary

In multiplication with tens and hundreds, we can use the multiplication table by thinking of sets of 10 and 100.

Want to confirm

 Let's calculate the following.

① 20 × 3 ② 30 × 3 ③ 80 × 2 ④ 50 × 6

⑤ 300 × 2 ⑥ 200 × 3 ⑦ 400 × 3 ⑧ 500 × 5

❷ How to multiply (2-digit number) × (1-digit number)

1   A child bought 3 sheets of colored paper which cost 23 yen each. How much was the total cost?

23 yen 23 yen 23 yen

① Let's write a math expression.

$$\boxed{} \times \boxed{}$$

② Approximately, how much is it?

If it costs 20 yen each ....

Yui

Want to think

③ Let's think about how to calculate.

Can I calculate by using the number of 10-yen and 100-yen coins?

Daiki

When we calculated 12 × 4, we decomposed 12 into 10 and 2.

Yui

Purpose How can we calculate 23 × 3?

Want to explain

④ Let's explain the ideas of the following two children.

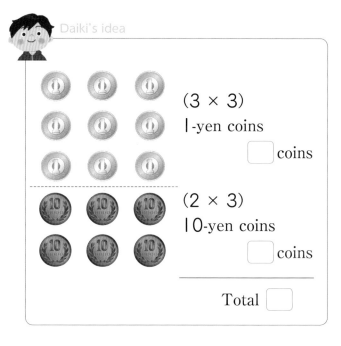

Daiki's idea

(3 × 3)
1-yen coins
☐ coins

(2 × 3)
10-yen coins
☐ coins

Total ☐

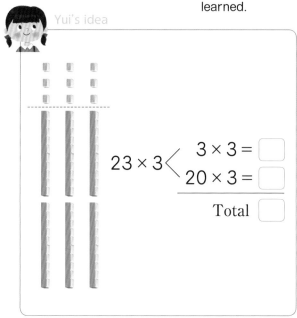

Yui's idea

$23 \times 3 \begin{cases} 3 \times 3 = \boxed{} \\ 20 \times 3 = \boxed{} \end{cases}$

Total ☐

Math Sentence : 23 × 3 = 69   Answer : 69 yen

Want to compare

⑤ Let's discuss the ideas they have in common.

⚥ Summary

When calculating 23 × 3, if we decompose 23 into 20 and 3, we can calculate by using the multiplication table.

$23 \times 3 \begin{cases} 3 \times 3 = 9 \\ 20 \times 3 = 60 \end{cases}$

Total 69

Want to connect

Can I calculate 23 × 3 in vertical form as I add and subtract?

Nanami

We can make multiplication algorithm in **vertical form**.

$$\begin{array}{r} 2\ 3 \\ \times\quad 3 \\ \hline \end{array}$$

2 Let's think about how to calculate 23 × 3 in vertical form.

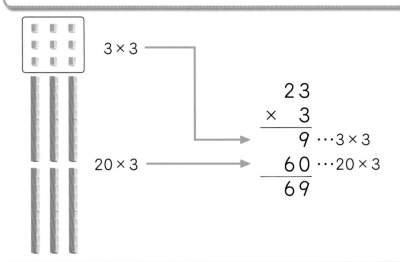

Multiplication algorithm for 23 × 3 in vertical form

$$\begin{array}{r} 2\ 3 \\ \times\quad 3 \\ \hline \end{array} \quad\rightarrow\quad \begin{array}{r} 2\ 3 \\ \times\quad 3 \\ \hline 9 \end{array} \quad\rightarrow\quad \begin{array}{r} 2\ 3 \\ \times\quad 3 \\ \hline 6\ 9 \end{array}$$

Align the digits of the numbers vertically according to their places.

3 × 3 is 9.
9 is in the ones place.

3 × 2 is 6.
6 is in the tens place.

1 Let's calculate the following in vertical form.

①
$$\begin{array}{r} 1\ 2 \\ \times\quad 4 \\ \hline \end{array}$$

②
$$\begin{array}{r} 3\ 3 \\ \times\quad 3 \\ \hline \end{array}$$

③
$$\begin{array}{r} 2\ 4 \\ \times\quad 2 \\ \hline \end{array}$$

 2 Let's calculate the following in vertical form.

① 34 × 2   ② 21 × 3   ③ 42 × 2   ④ 11 × 4

3 Let's think about how to multiply in vertical form.

Carrying to the hundreds place

① 71 × 4

4 × 1 is 4.
☐ is in the ones place.

4 × 7 is 28.
8 is in the tens place.
☐ is in the hundreds place.

28 means 28 sets of what?

Daiki

Carrying to the tens place

② 13 × 7

7 × 3 is 21.
1 is in the ones place.
2 is carried to the tens place.

7 × 1 is 7.
Add the carried 2 to 7.
7 + 2 = ☐
☐ is in the tens place.

Carrying to the tens place and the hundreds place ①

③ 95 × 3

3 × 5 is 15.
☐ is in the ones place.
1 is carried to the tens place.

3 × 9 is 27.
Add the carried 1 to 27.
27 + 1 = ☐
☐ is in the tens place.
☐ is in the hundreds place.

Carrying to the tens place and the hundreds place ②

④ 46 × 7

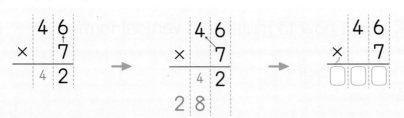

After multiplying in the tens place, be careful of carrying in the addition.

7 × 6 is 42.

2 is in the ones place.

☐ is carried to the tens place.

7 × 4 is 28.

Add the carried 4 to 28.

28 + 4 = ☐

☐ is in the tens place.

☐ is in the hundreds place.

Want to confirm

3 Let's calculate the following in vertical form.

① 93 × 3　② 63 × 2　③ 51 × 3　④ 30 × 8

⑤ 14 × 7　⑥ 13 × 5　⑦ 64 × 3　⑧ 47 × 4

⑨ 19 × 7　⑩ 38 × 3　⑪ 18 × 6　⑫ 25 × 4

⑬ 59 × 7　⑭ 35 × 9　⑮ 65 × 8　⑯ 84 × 6

4 You bought 4 snacks that cost 55 yen each. How much was the total cost?

5 Each problem has a letter. If you arrange the answers to each problem from smallest to largest, what is the hidden phrase?

| Ⓐ 73×8 | Ⓗ 87×9 | Ⓣ 93×8 | Ⓞ 68×4 | Ⓛ 30×9 |
| Ⓜ 57×8 | Ⓔ 42×9 | Ⓦ 12×8 | Ⓥ 46×6 | Ⓔ 31×5 |

Want to know

Activity

1 You bought a 3 m ribbon that costs 213 yen per meter. How much was the cost?

① Let's write a math expression.

② Approximately, how much is it?

If it costs 200 yen per meter ....

Yui

Want to think

③ Let's think about how to calculate.

Purpose How can we calculate 213 × 3?

Hiroto's idea

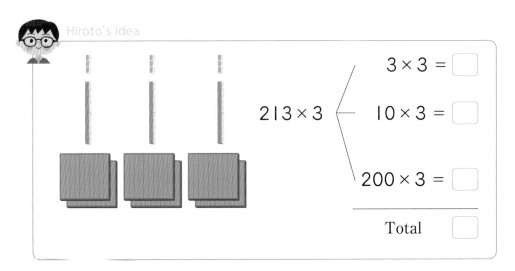

213×3
$3 \times 3 = \boxed{}$
$10 \times 3 = \boxed{}$
$200 \times 3 = \boxed{}$
Total $\boxed{}$

Way to see and think

When you calculated 23 × 3, you decomposed 23 into 20 and 3.

Way to see and think

In 200 × 3, he uses multiplication with hundreds.

Math sentence : $213 \times 3 = 639$ Answer : 639 yen

Summary

When calculating 213 × 3, first decompose 213 into 200, 10, and 3, then use the multiplication table.

213×3
$3 \times 3 = 9$
$10 \times 3 = 30$
$200 \times 3 = 600$
Total 639

Want to connect

Can I calculate 213 × 3 in vertical form?

Daiki

2 Let's think about how to calculate 213×3.

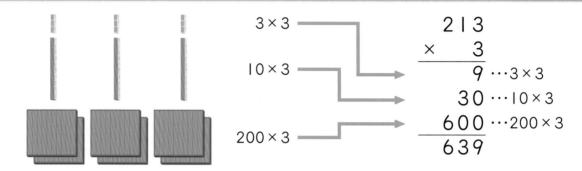

3×3
10×3
200×3

$$\begin{array}{r} 213 \\ \times \quad 3 \\ \hline 9 \cdots 3 \times 3 \\ 30 \cdots 10 \times 3 \\ 600 \cdots 200 \times 3 \\ \hline 639 \end{array}$$

Multiplication algorithm for 213×3 in vertical form

3×3 is 9.
9 is in the ones place.

→

3×1 is 3.
3 is in the tens place.

→

3×2 is 6.
6 is in the hundreds place.

Multiplication with carrying

 Let's explain how to calculate 461×3 in vertical form.

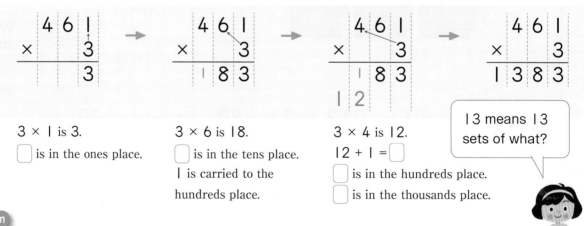

3×1 is 3.
◯ is in the ones place.

3×6 is 18.
◯ is in the tens place.
1 is carried to the hundreds place.

3×4 is 12.
$12 + 1 = $ ◯
◯ is in the hundreds place.
◯ is in the thousands place.

13 means 13 sets of what?

Nanami

 Let's calculate the following in vertical form.

① 142×2 ② 423×2 ③ 312×3 ④ 121×4

⑤ 321×4 ⑥ 341×5 ⑦ 427×3 ⑧ 819×2

3 Let's explain how to calculate the following in vertical form.

① 876 × 7

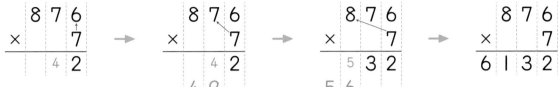

7 × 6 is 42.
☐ is in the ones place.
4 is carried to the tens place.

7 × 7 is 49.
49 + 4 = ☐
☐ is in the tens place.
5 is carried to the hundreds place.

7 × 8 is 56.
56 + 5 = ☐
☐ is in the hundreds place.
☐ is in the thousands place.

② 334 × 3

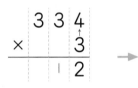

3 × 4 is 12.
☐ is in the ones place.
1 is carried to the tens place.

3 × 3 is 9.
9 + 1 = ☐
☐ is in the tens place.
1 is carried to the hundreds place.

3 × 3 is 9.
9 + 1 = ☐
☐ is in the hundreds place.
☐ is in the thousands place.

3 Let's calculate the following in vertical form.

① 654 × 3   ② 364 × 8   ③ 749 × 7
④ 128 × 8   ⑤ 429 × 7   ⑥ 367 × 9

4 You bought 3 sandwiches that cost 238 yen each. How much was the total cost?

4 Let's calculate 405×8 in vertical form.

```
  4 0 5
×     8
------
  4 0
```
→
```
  4 0 5
×     8
------
3 2 4 0
```

```
    4 0 5
×       8
--------
      4 0  ···5 × 8
      0 0  ···0 × 8
    3 2 0 0 ···400 × 8
--------
    3 2 4 0
```

5 Let's calculate the following in vertical form.

① 302×9　② 505×6　③ 420×7　④ 600×2

4 Mental calculation

1 You bought 3 candies that cost 24 yen each. How much was the total cost?

Let's try to find the answer mentally.

Nanami

I remembered how to multiply in vertical form.

24×3 (1) (2)

(1) 3 × 4 is 12.
(2) 3 × 2 is 6, so it's 60.
(3) 12 + 60 = 72

Daiki

I multiplied from the tens place.

24×3 (1) (2)

(1) 3 × 2 is 6, so it's 60.
(2) 3 × 4 is 12.
(3) 60 + 12 = 72

It is easier to find the approximate cost with Daiki's method.

1 Let's calculate 63×4 mentally.

2 Let's calculate the following mentally.

① 34×2　② 17×3　③ 25×4　④ 44×4

What you can do now

☐ Understanding how to multiply in vertical form.

1 You explained how to calculate 48 × 6 in vertical form.

Let's fill in the ☐ with numbers.

```
  48
×  6
─────
```

[in the ones place] 6 × 8 is 48

☐ is in the ones place.

☐ is carried to the tens place.

[in the tens place] 6 × 4 is 24

24 + 4 = ☐

☐ is in the tens place.

☐ is in the hundreds place.

☐ Can multiply by 1-digit number in vertical form.

2 Let's calculate the following in vertical form.

① 50 × 3   ② 300 × 3

③ 600 × 7   ④ 22 × 4

⑤ 38 × 2   ⑥ 64 × 8

⑦ 94 × 6   ⑧ 223 × 3

⑨ 179 × 8   ⑩ 379 × 7

⑪ 584 × 5   ⑫ 715 × 7

☐ Can make a multiplication expression and find the answer.

3 Let's answer the following.

① There is a park near Reina's house. The road distance around the park is 340 m. She ran around the park 4 times. How many meters did she run in all?

② You bought 6 goldfish that cost 125 yen each. How much was the total cost?

Supplementary problems ▶ p.137

Usefulness and efficiency of learning

1 Let's find the mistakes in each and write the correct answer.

①
```
    5 8
  ×   6
  3 0 4 8
```
(   )

②
```
    2 7 6
  ×     4
    8 0 4
```
(   )

③
```
    6 1 5
  ×     7
  4 2 0 5
```
(   )

Understanding how to multiply in vertical form.

2 Let's calculate in vertical form.

① 20 × 8
② 900 × 5
③ 400 × 6
④ 56 × 4
⑤ 76 × 8
⑥ 43 × 7
⑦ 324 × 2
⑧ 254 × 6
⑨ 483 × 5
⑩ 408 × 9
⑪ 112 × 9
⑫ 638 × 8

Can multiply by 1-digit number in vertical form.

3 Let's answer the following.

① A notebook costs 125 yen and a pencil costs 40 yen. If you buy 8 sets of notebooks and pencils, how much is the total cost?

② You bought 4 bags of French fries that cost 35 yen each and a chocolate that costs 108 yen. If you gave 1000 yen, how much was the change?

Can make a multiplication expression and find the answer.

French Fries

Chocolate

4 Let's make a math problem for 43 × 6.

Can make a math problem for a multiplication expression.

126

❶ Multiplication

pp.10~22

❙ Let's fill in the ☐ with numbers.

① $7 \times 3 = 3 \times \boxed{}$

② $8 \times 6 = \boxed{} \times 8$

❷ Let's fill in the ☐ with numbers.

① 6×5 is larger than 6×4 by $\boxed{}$.

② 9×4 is larger than $9 \times \boxed{}$ by 9.

③ 8×7 is smaller than 8×8 by $\boxed{}$.

④ 3×5 is smaller than $3 \times \boxed{}$ by 3.

⑤ $4 \times 9 = 4 \times \boxed{} + 4$

⑥ $7 \times 6 = 7 \times 5 + \boxed{}$

⑦ $5 \times 7 = 5 \times 8 - \boxed{}$

⑧ $9 \times 8 = 9 \times \boxed{} - 9$

❸ Let's fill in the ☐ with numbers.

① 8×5 ⟨ $3 \times 5 = \boxed{}$

$\boxed{} \times 5 = \boxed{}$

Total $\boxed{}$

② 9×7 ⟨ $9 \times 2 = \boxed{}$

$9 \times \boxed{} = \boxed{}$

Total $\boxed{}$

❹ Let's fill in the ☐ with numbers.

① $(3 \times 2) \times 2 = \boxed{} \times 2$

$= \boxed{}$

$3 \times (2 \times 2) = 3 \times \boxed{}$

$= \boxed{}$

$(3 \times 2) \times 2 = \boxed{} \times (2 \times 2)$

② $(3 \times 2) \times 3 = 3 \times (2 \times \boxed{})$

③ $(4 \times 2) \times 2 = 4 \times (\boxed{} \times 2)$

④ $(2 \times 3) \times 2 = \boxed{} \times (3 \times 2)$

❺ Let's calculate the following.

① 2×0 ② 7×0

③ 0×5 ④ 9×0

⑤ 0×3 ⑥ 0×8

❻ Let's calculate the following.

① 3×10 ② 7×10

③ 9×10 ④ 10×6

⑤ 10×8 ⑥ 10×5

❼ There are 4 boxes of cookies that contain 10 cookies each. How many cookies are there altogether?

② Time and Duration(1)

pp.23～31

1 Let's find the following times.

① The time 30 minutes past 8:50 a.m.

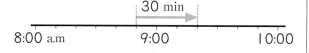

② The time 50 minutes past 3:40 p.m.

③ The time 1 hour 20 minutes past 10:30 a.m.

④ The time 2 hours 20 minutes past 6:45 p.m.

2 Let's find the following durations.

① The duration from 8:40 a.m. to 9:20 a.m.

② The duration from 1:50 p.m. to 2:40 p.m.

③ The duration from 9:30 a.m. to 11:10 a.m.

④ The duration from 4:45 p.m. to 7:15 p.m.

3 Let's find the following times.

① The time 30 minutes before 11:10 a.m.

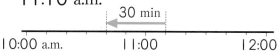

② The time 50 minutes before 4:30 p.m.

③ The time 1 hour 40 minutes before 9:40 a.m.

④ The time 2 hours 30 minutes before 5:15 p.m.

4 Let's calculate the following times and durations.

① The time 30 minutes before 8:50

Hr	Min
8	50
−	30

② The duration from 1:20 to 6:50

Hr	Min
6	50
− 1	20

③ The time 40 minutes past 1:40

Hr	Min
1	40
+	40

④ The time 1 hour 40 minutes past 4:10

Hr	Min
4	10
+ 1	40

③ Division

pp.32〜47

① If you divide 18 chocolates equally among 3 children, how many chocolates does each child get? Let's write a math expression.

② If you divide 15 dL of milk equally into 5 cups, how many dL of milk does each cup have? Let's write a math expression.

③ Let's fill in the ☐ with numbers.
   The answer to 20 ÷ 4 can be found by using the row of ☐ in the multiplication table.

$4 \times 2 = 8$

$4 \times 3 = 12$

$4 \times 4 = \boxed{}$

$4 \times 5 = \boxed{}$

So, $20 \div 4 = \boxed{}$.

④ Which row in the multiplication table can be used to find the answer to each division? Let's find the answer.
   ① $4 \div 2$ ② $28 \div 7$
   ③ $56 \div 8$ ④ $24 \div 4$
   ⑤ $48 \div 6$ ⑥ $63 \div 9$

⑤ You are going to make a math problem for each expression.
Let's fill in the ☐ with numbers.
   ① $24 \div 3$

   If you divide ☐ sheets of colored paper equally among ☐ children, how many sheets does each child get?

   ② $30 \div 5$

   If you cut a ☐ cm tape equally into ☐ , how long is each tape?

⑥ Let's calculate the following.
   ① $16 \div 8$ ② $14 \div 2$
   ③ $27 \div 3$ ④ $35 \div 5$
   ⑤ $24 \div 6$ ⑥ $28 \div 4$

⑦ There are 18 candies and each child gets 3 candies. Let's think about how many children can share the candies.
   ① Let's write a math expression.
   ② Which row in the multiplication table can be used to find the answer to the division you made in ①?
   ③ How many children can share the candies?

8 There are 20 dL of juice. If you pour 4 dL into each bottle, how many bottles will be filled?

9 Let's calculate the following.
① $14 \div 7$ ② $18 \div 9$
③ $25 \div 5$ ④ $36 \div 4$
⑤ $21 \div 3$ ⑥ $16 \div 2$
⑦ $56 \div 7$ ⑧ $42 \div 6$
⑨ $54 \div 9$ ⑩ $72 \div 8$

10 You are going to make a math problem for $27 \div 9$. Let's fill in the ☐ with numbers.

┌─────────────────────────────────┐
There are ☐ cookies. If each child gets ☐ cookies, how many children can share the cookies?
└─────────────────────────────────┘

11 There are 30 cards. Let's think about how to share them.
① If 5 children get the same number of cards, how many cards does each child get?
② If each child gets 6 cards, how many children can share the cards?

12 Let's think about how to cut a 24 cm ribbon.
① If you cut it every 6 cm, how many ribbons can you get?
② If you cut it equally into 3, how long is each ribbon?

13 Let's calculate the following.
① $8 \div 8$ ② $3 \div 3$
③ $0 \div 7$ ④ $0 \div 5$
⑤ $4 \div 1$ ⑥ $8 \div 1$

14 Let's calculate the following.
① $20 \div 2$ ② $60 \div 3$
③ $44 \div 4$ ④ $22 \div 2$

④ Division with Remainders
pp.48~59

11 Let's think about how to divide 18 candies so that each child get 4 candies.
① Can 5 children get 4 candies?
② Let's fill in the ☐ with numbers.
$18 \div 4 = $ ☐ remainder ☐
③ How many children can get 4 candies and how many candies will remain?

2 There are 33 pencils. The same number of pencils will be given to 6 children. How many pencils will each child get? How many pencils will remain?

3 Let's confirm the answers of the following. If the answer is not correct, write the correct answer.
① $30 \div 4 = 8$ remainder 2
② $48 \div 7 = 7$ remainder 1
③ $29 \div 5 = 5$ remainder 4

4 You are going to calculate the following and confirm the answers. Let's fill in the ☐ with numbers.
① $17 \div 2 = \boxed{}$ remainder $\boxed{}$
confirmation
$2 \times \boxed{} + \boxed{} = \boxed{}$
② $43 \div 8 = \boxed{}$ remainder $\boxed{}$
confirmation
$8 \times \boxed{} + \boxed{} = \boxed{}$

5 Let's calculate the following and confirm the answers.
① $9 \div 4$ ② $8 \div 3$
③ $46 \div 5$ ④ $29 \div 6$
⑤ $40 \div 9$ ⑥ $59 \div 7$

6 Let's think about how to distribute 46 sheets of drawing paper.
① If each child gets 6 sheets, how many children can share the drawing paper?
How many sheets will remain?
② If 8 children get the same number of sheets, how many sheets does each one get?
How many sheets will remain?

7 Let's think about how to cut a 50 cm ribbon.
① If you cut it every 7 cm, how many ribbons can you get? How many cm will remain?
② If you want to get 8 ribbons of 7 cm, how many more cm of ribbon do you need?

8 There are 60 cakes that are to be placed on plates. If 8 cakes can be placed on each plate, how many plates are needed?

9 You are going to make packs of 6 eggs. There are 40 eggs. How many packs will be filled?

⑤ Addition and Subtraction

pp.62~81

① Let's calculate the following in vertical form.

① $352 + 416$ ② $316 + 253$

③ $648 + 151$ ④ $652 + 107$

⑤ $108 + 471$ ⑥ $306 + 401$

② Let's calculate the following in vertical form.

① $128 + 433$ ② $516 + 248$

③ $367 + 527$ ④ $247 + 236$

⑤ $678 + 119$ ⑥ $365 + 308$

③ Let's calculate the following in vertical form.

① $362 + 451$ ② $671 + 275$

③ $240 + 380$ ④ $693 + 237$

⑤ $189 + 442$ ⑥ $736 + 89$

④ Let's calculate the following in vertical form.

① $273 + 229$ ② $415 + 387$

③ $532 + 369$ ④ $656 + 144$

⑤ $488 + 312$ ⑥ $334 + 68$

⑤ There are 257 sheets of red paper and 163 sheets of blue paper. How many sheets of paper are there altogether?

⑥ Let's calculate the following in vertical form.

① $873 - 241$ ② $659 - 448$

③ $679 - 565$ ④ $689 - 525$

⑤ $576 - 256$ ⑥ $816 - 511$

⑦ Let's calculate the following in vertical form.

① $256 - 138$ ② $674 - 405$

③ $630 - 215$ ④ $816 - 332$

⑤ $345 - 263$ ⑥ $718 - 631$

⑧ Let's calculate the following in vertical form.

① $556 - 278$ ② $322 - 199$

③ $934 - 289$ ④ $311 - 163$

⑤ $340 - 165$ ⑥ $614 - 58$

⑨ Let's calculate the following in vertical form.

① $504 - 346$ ② $906 - 438$

③ $804 - 459$ ④ $604 - 206$

⑤ $200 - 153$ ⑥ $500 - 45$

⑩ Yuri has read 145 pages of a book with 240 pages. How many pages does she still need to read?

11 Let's calculate the following in vertical form.

① $685 + 536$ ② $483 + 517$

③ $1327 - 943$ ④ $1007 - 979$

⑤ $3496 + 4207$

⑥ $7382 + 2618$

⑦ $7134 - 3146$

⑧ $7225 - 4627$

⑨ $10000 - 3027$

12 Let's fill in the ☐ with numbers.

① $258 + 98 = 258 + \boxed{} - 2$

$= \boxed{} - 2$

$= \boxed{}$

② $406 - 197 = 406 - \boxed{} + 3$

$= \boxed{} + 3$

$= \boxed{}$

③ $157 + 76 + 24$

$= 157 + (\boxed{} + 24)$

$= 157 + \boxed{}$

$= \boxed{}$

13 Let's calculate the following in easier ways.

① $199 + 165$

② $302 - 98$

③ $277 + 68 + 32$

④ $43 + 188 + 57$

14 Let's calculate the following mentally.

① $17 + 8$ ② $56 + 16$

③ $26 - 9$ ④ $82 - 24$

15 There are 416 children in Hiroka's school. 198 of them are boys. How many girls are there in her school?

16 There are 165 red cosmoses that bloomed. The number of white cosmoses that bloomed is 15 more than that of red cosmoses. How many white cosmoses bloomed?

6 Tables and Graphs
pp.84~97

11 The table below shows the data about the vehicles that passed in front of the school. Let's change the character "正" to numbers. Then, let's find the total number.

Vehicles that passed in front of the school

Kind	Number of vehicles	
Truck	正	
Car	正 下	
Bus	丁	
Others	下	
Total		

2 The graph below shows the data about vehicles that passed in front of the station.

① How many vehicles does 1 cell represent?

Vehicles that passed in front of the station

(vehicles)

② What kind of vehicle passed most frequently? How many were there?

③ How many trucks and buses were there respectively?

3 The bar graph below shows the number of empty cans collected by the children in each grade.

Empty cans collected by children

0   100   200   300 (cans)

Grade 1
Grade 2
Grade 3
Grade 4
Grade 5
Grade 6

① How many cans does 1 cell represent?

② What grade collected the most cans? How many did they collect?

③ Which grade collected more cans, 3rd or 6th, and by how many?

4 In the graphs below, let's tell the number that each bar represents.

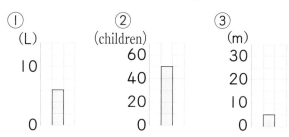

5 The table below shows the favorite fruits of children in Daichi's class. Let's draw a bar graph.

Favorite fruits   (children)

Apple	Grape	Strawberry	Orange	Others
12	9	6	5	4

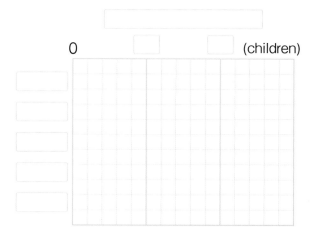

6   The following tables show the number of the 3rd grade children who borrowed each kind of books in April, May, and June. Let's organize the tables for each month into one table.

April

Kind	Number of Books
Fiction	13
Biography	5
Picture	7
Others	4

May

Kind	Number of Books
Fiction	18
Biography	12
Picture	9
Others	6

June

Kind	Number of Books
Fiction	17
Biography	11
Picture	13
Others	8

Books borrowed by the 3rd grade children   (books)

Kind \ Month	April	May	June	Total
Fiction				
Biography				
Picture				
Others				
Total				

7 Length

pp.100~110

1   How many m and cm are the lengths of points, ①, ②, ③, and ④ located on the tape measure below?

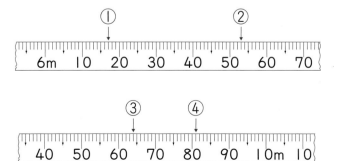

2   Let's fill in the ☐ with numbers.

①   4000 m = ☐ km

②   2850 m = ☐ km ☐ m

③   7 km = ☐ m

④   3 km 60 m = ☐ m

3   Which is longer?

①   1 km 690 m, 1800 m

②   3 km 75 m, 3065 m

4   Look at the map below and answer the following.

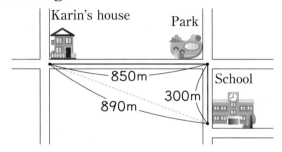

①   How many km and m is the road distance from Karin's house to the school through the front of the park?

②   What is the difference between the road distance and the distance from Karin's house to the school in m?

5   Let's calculate the following.

①   2 km 750 m + 1 km 240 m

②   2 km 380 m + 1 km 640 m

③   2 km 480 m − 1 km 260 m

④   3 km 80 m − 1 km 630 m

9 Multiplication with 1-digit Number

pp.114~126

1 Let's calculate the following.
① 30 × 2   ② 40 × 4
③ 50 × 8   ④ 60 × 9

2 Let's calculate the following.
① 400 × 2   ② 300 × 6
③ 800 × 4   ④ 500 × 8

3 Let's fill in the ☐ with numbers.

Ones place   Tens place

```
   3 2              3 2
 ×   2            ×   2
 ┌───┐           ┌───┐
 │   │           │   │ 4
 └───┘           └───┘
```

4 Let's calculate the following in vertical form.
① 23 × 2   ② 22 × 2
③ 13 × 3   ④ 43 × 2

5 Let's calculate the following in vertical form.
① 42 × 3   ② 94 × 2
③ 31 × 5   ④ 53 × 2
⑤ 24 × 3   ⑥ 49 × 2
⑦ 16 × 6   ⑧ 87 × 3
⑨ 46 × 4   ⑩ 35 × 5

6 You bought 3 chocolates that cost 45 yen each.

How much was the total cost?

7 Let's calculate the following in vertical form.
① 27 × 8   ② 89 × 6
③ 78 × 4   ④ 28 × 9
⑤ 48 × 7   ⑥ 87 × 6
⑦ 19 × 6   ⑧ 75 × 4

8 Let's calculate the following in vertical form.
① 132 × 2   ② 322 × 3
③ 316 × 4   ④ 964 × 2
⑤ 751 × 6   ⑥ 261 × 7

9 Let's calculate the following in vertical form.
① 748 × 3   ② 846 × 4
③ 667 × 8   ④ 347 × 7
⑤ 673 × 3   ⑥ 835 × 6
⑦ 556 × 9   ⑧ 467 × 6
⑨ 370 × 4   ⑩ 605 × 4

10 You bought 6 cakes that cost 175 yen each. How much is the total cost?

Answers

1 Multiplication

1. ① 7 ② 6
2. ① 6 ② 3 ③ 8 ④ 6 ⑤ 8 ⑥ 7 ⑦ 5
   ⑧ 9
3. ① 8×5 < $3 \times 5 = \boxed{15}$
                  $\boxed{5} \times 5 = \boxed{25}$
   Total $\boxed{40}$
   ② 9×7 < $9 \times 2 = \boxed{18}$
                  $9 \times \boxed{5} = \boxed{45}$
   Total $\boxed{63}$
4. ① 6, 12, 4, 12, 3 ② 3 ③ 2 ④ 2
5. ① 0 ② 0 ③ 0 ④ 0 ⑤ 0 ⑥ 0
6. ① 30 ② 70 ③ 90 ④ 60 ⑤ 80 ⑥ 50
7. $10 \times 4 = 40$ 40 cookies

2 Time and Duration(1)

1. ① 9:20 a.m. ② 4:30 p.m. ③ 11:50 a.m.
   ④ 9:05 p.m.
2. ① 40 minutes ② 50 minutes ③ 1 hour 40 minutes
   ④ 2 hours 30 minutes
3. ① 10:40 a.m. ② 3:40 p.m. ③ 8:00 a.m.
   ④ 2:45 p.m.
4. ① 8:20 ② 5 hours 30 minutes ③ 2:20 ④ 5:50

3 Division

1. $18 \div 3$
2. $15 \div 5$
3. 4, 16, 20, 5
4. ① row of 2, 2 ② row of 7, 4 ③ row of 8, 7
   ④ row of 4, 6 ⑤ row of 6, 8 ⑥ row of 9, 7
5. ① 24, 3 ② 30, 5
6. ① 2 ② 7 ③ 9 ④ 7 ⑤ 4 ⑥ 7
7. ① $18 \div 3$ ② row of 3 ③ 6 children
8. $20 \div 4 = 5$ 5 bottles
9. ① 2 ② 2 ③ 5 ④ 9 ⑤ 7 ⑥ 8 ⑦ 8
   ⑧ 7 ⑨ 6 ⑩ 9
10. 27, 9
11. ① $30 \div 5 = 6$ 6 cards ② $30 \div 6 = 5$ 5 children
12. ① $24 \div 6 = 4$ 4 ribbons ② $24 \div 3 = 8$ 8 cm
13. ① 1 ② 1 ③ 0 ④ 0 ⑤ 4 ⑥ 8
14. ① 10 ② 20 ③ 11 ④ 11

4 Division with Remainders

1. ① No. ② 4, 2
   ③ 4 children; 2 candies remain
2. $33 \div 6 = 5$ remainder 3   5 pencils; 3 pencils remain
3. ① 7 remainder 2 ② 6 remainder 6 ③ ○
4. ① 8, 1, 8, 1, 17 ② 5, 3, 5, 3, 43
5. ① 2 remainder 1, $4 \times 2 + 1 = 9$
   ② 2 remainder 2, $3 \times 2 + 2 = 8$
   ③ 9 remainder 1, $5 \times 9 + 1 = 46$
   ④ 4 remainder 5, $6 \times 4 + 5 = 29$
   ⑤ 4 remainder 4, $9 \times 4 + 4 = 40$
   ⑥ 8 remainder 3, $7 \times 8 + 3 = 59$
6. ① $46 \div 6 = 7$ remainder 4   7 children; 4 sheets remain
   ② $46 \div 8 = 5$ remainder 6
   5 sheets for each child; 6 sheets remain
7. ① $50 \div 7 = 7$ remainder 1   7 ribbons; 1cm remains
   ② $7 \times 8 = 56$ $56 - 50 = 6$ 6 cm
8. $60 \div 8 = 7$ remainder 4   8 plates
9. $40 \div 6 = 6$ remainder 4   6 packs

5 Addition and Subtraction

1. ① 768 ② 569 ③ 799 ④ 759 ⑤ 579 ⑥ 707
2. ① 561 ② 764 ③ 894 ④ 483 ⑤ 797 ⑥ 673
3. ① 813 ② 946 ③ 620 ④ 930 ⑤ 631 ⑥ 825
4. ① 502 ② 802 ③ 901 ④ 800 ⑤ 800 ⑥ 402
5. $257 + 163 = 420$ 420 sheets
6. ① 632 ② 211 ③ 114 ④ 164 ⑤ 320 ⑥ 305
7. ① 118 ② 269 ③ 415 ④ 484 ⑤ 82 ⑥ 87
8. ① 278 ② 123 ③ 645 ④ 148 ⑤ 175 ⑥ 556
9. ① 158 ② 468 ③ 345 ④ 398 ⑤ 47 ⑥ 455
10. $240 - 145 = 95$ 95 pages
11. ① 1221 ② 1000 ③ 384 ④ 28 ⑤ 7703
    ⑥ 10000 ⑦ 3988 ⑧ 2598 ⑨ 6973
12. ① 100, 358, 356 ② 200, 206, 209
    ③ 76, 100, 257
13. ① 364 ② 204 ③ 377 ④ 288
14. ① 25 ② 72 ③ 17 ④ 58
15. $416 - 198 = 218$ 218 girls
16. $165 + 15 = 180$ 180 white cosmoses

6 Tables and Graphs

1 Vehicles that passed in front of the school

Kind		Number of vehicles
Truck	正	5
Car	正下	8
Bus	丅	2
Others	下	3
Total		18

2 ① 1 vehicle ② 12 cars

③ 3 trucks 9 buses

3 ① 20 cans ② 4th grade, 260 cans

③ 6th grade, 20 cans

4 ① 6 L ② 50 children ③ 5 m

5

Favorite fruits

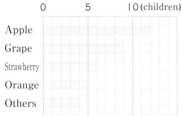

6 Books borrowed by the 3rd grade children (books)

Month Kind	April	May	June	Total
Fiction	13	18	17	48
Biography	5	12	11	28
Picture	7	9	13	29
Others	4	6	8	18
Total	29	45	49	123

7 Length

1 ① 6m17cm ② 6m53cm

③ 9m64cm ④ 9m81cm

2 ① 4 ② 2, 850 ③ 7000 ④ 3060

3 ① 1800m ② 3km75m

4 ① 1km150m ② 260m

5 ① 3km990m ② 4km20m ③ 1km220m

④ 1km450m

9 Multiplication with 1-digit Numbers

1 ① 60 ② 160 ③ 400 ④ 540

2 ① 800 ② 1800 ③ 3200 ④ 4000

3 4, 6

4 ① 46 ② 44 ③ 39 ④ 86

5 ① 126 ② 188 ③ 155 ④ 106 ⑤ 72

⑥ 98 ⑦ 96 ⑧ 261 ⑨ 184 ⑩ 175

6 45 × 3 = 135   135 yen

7 ① 216 ② 534 ③ 312 ④ 252

⑤ 336 ⑥ 522 ⑦ 114 ⑧ 300

8 ① 264 ② 966 ③ 1264

④ 1928 ⑤ 4506 ⑥ 1827

9 ① 2244 ② 3384 ③ 5336 ④ 2429

⑤ 2019 ⑥ 5010 ⑦ 5004 ⑧ 2802

⑨ 1480 ⑩ 2420

10 175 × 6 = 1050   1050 yen

Symbols and words in this book

Multiplication Table

▼ will be used in pages 11~20.

Multiplier

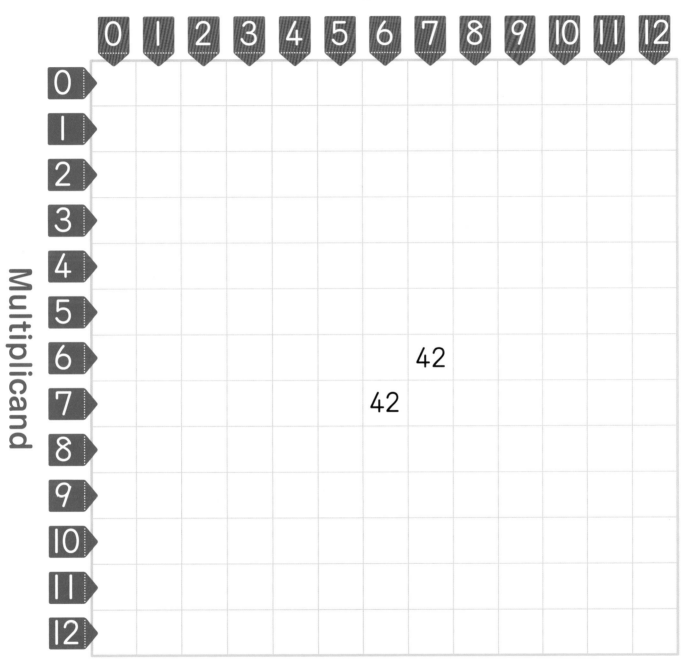

Multiplicand

	0	1	2	3	4	5	6	7	8	9	10	11	12
0													
1													
2													
3													
4													
5													
6								42					
7							42						
8													
9													
10													
11													
12													

Let's make the rows of 10, 11, and 12.

Nanami

Memo

Editorial for English Edition:

Study with Your Friends, Mathematics for Elementary School

3rd Grade, Vol.1, Gakko Tosho Co.,Ltd., Tokyo, Japan [2020]

Chief Editor:

Masami Isoda (University of Tsukuba, Japan), Aki Murata (University of Florida, USA)

Advisory Board:

Abrham Arcavi (Weizmann Institute of Science, Israel), Aida Istino Yap (University of the Philippines, Philippines), Alf Coles (University of Bristol, UK), Bundit Thipakorn (King Mongkut's University of Technology Thonburi, Thailand), Fou-Lai Lin (National Taiwan Normal University, Taiwan), Hee-Chan Lew (Korean National University of Education, Korea), Lambas (Ministry of Education and Culture, Indonesia), Luc Trouche (Ecole Normale Supérieure de Lyon, France), Maitree Inprasitha (Khon Kaen University, Thailand), Marcela Santillán (Universidad Pedagógica Nacional, Mexico), María Jesús Honorato Errázuriz (Ministry of Education, Chile), Raimundo Olfos Ayarza (Pontificia Universidad Católica de Valparaíso, Chile), Rogin Huang (Middle Tennessee State University, USA), Suhaidah Binti Tahir (SEAMEO RECSAM, Malaysia), Sumardyono (SEAMEO QITEP in Mathematics, Indonesia), Toh Tin Lam (National Institute of Education, Singapore), Toshikazu Ikeda (Yokohama National University, Japan), Wahyudi (SEAMEO Secretariat, Thailand), Yuriko Yamamoto Baldin (Universidade Federal de São Carlos, Brazil)

Editorial Board:

Abolfazl Rafiepour (Shahid Bahonar University of Kerman, Iran), Akio Matsuzaki (Saitama University, Japan), Cristina Esteley (Universidad Nacional de Córdoba, Argentina), Guillermo P. Bautista Jr. (University of the Philippines, Philippines), Ivan Vysotsky (Moscow Center for Teaching Excellence, Russia), Kim-Hong Teh (SEAMEO RECSAM, Malaysia), María Soledad Estrella (Pontificia Universidad Católica de Valparaiso, Chile), Narumon Changsri (Khon Kaen University, Thailand), Nguyen Chi Thanh (Vietnam National University, Vietnam), Onofre Jr Gregorio Inocencio (Foundation to Upgrade the Standard of Education, Philippines), Ornella Robutti (Università degli Studi di Torino, Iraly), Raewyn Eden (Massey University, New Zealand), Roberto Araya (Universidad de Chile, Chile), Soledad Asuncion Ulep (University of the Philippines, Philippines), Steven Tandale (Department of Education, Papua New Guinea), Tadayuki Kishimoto (Toyama University, Japan), Takeshi Miyakawa (Waseda University, Japan), Tenoch Cedillo (Universidad Pedagógica Nacional, Mexico), Ui Hock Cheah (IPG Pulau Pinang, Malaysia), Uldarico Victor Malaspina Jurado (Pontificia Universidad Católica del Perú, Peru), Wahid Yunianto (SEAMEO QITEP in Mathematics, Indonesia), Wanty Widjaja (Deakin University, Australia)

Translators:

Masami Isoda, Solis Worsfold Diego, Tsuyoshi Nomura (University of Tsukuba, Japan)

Hideo Watanabe (Musashino University, Japan)